North Carolina 26th Regiment Infantry Roster and History

John W. Moore
and
Walter Clark

Heritage Books
2026

HERITAGE BOOKS
AN IMPRINT OF HERITAGE BOOKS, INC.

Books, CDs, and more—Worldwide

For our listing of thousands of titles see our website
at
www.HeritageBooks.com

A Facsimile Reprint
Published 2026 by
HERITAGE BOOKS, INC.
Publishing Division
5810 Ruatan Street
Berwyn Heights, MD 20740

Reprinted 2019
Mountain Press
Signal Mountain, Tennessee

International Standard Book Number
Paperbound: 978-0-7884-9251-8

TWENTY-SIXTH REGIMENT—INFANTRY.

ABBREVIATIONS.

c	captured.	d	died.	p	promoted.
cm	commissioned.	dg.	discharged.	pr	prisoner.
co	county.	dt	detailed.	r	resigned.
Com	Company.	k	killed.	tr	transferred.
e	enlisted.	m	missing.	w	wounded.

FIELD AND STAFF.

Zebulon B. Vance, Colonel, cm. August 27th, '61; Buncombe co.; p. Governor of North Carolina and Commander-in-Chief September 8th, '62.

Henry K. Burgwyn, Colonel, cm. in September, '62; Northampton co.; p. from Lieutenant Colonel, and k. July 1st, '63, at Gettysburg.

John R. Lane, Colonel, cm. July 1st, '63; Chatham co.; p. from Lieut. Colonel; w. at Gettysburg and Wilderness.

Henry K. Burgwyn, Lieutenant Colonel, cm. August 27th, '61; Northampton co.; p. and k.

John R. Lane, Lieutenant Colonel, cm. Sept., '62; Chatham co.; p. from Major; w. at New Bern.

T. J. Adams, Lieutenant Colonel, cm. July 1st, '63; p. from Major; w. at New Bern and Gettysburg.

Abner B. Carmichael, Major, cm. July 27th, '62; Wilkes co.; p. from Captain of Company C; k. at New Bern March 14th, '62.

John R. Lane, Major, cm. April 12th, '62; Chatham co.; p. from Captain of Company H.

Nathaniel P. Rankin, Major, cm. March 14th, '62; Guilford co.; p. from Captain of Company F.

John T. Jones, Major, cm. Sept., '62; Caldwell co.; p. from Captain of Com. I; k. May 6th, '64, at Wilderness.

T. J. Adams, Major, cm. in May, '64; p. from Captain of Company D.

James B. Jordan, Adjutant, cm. August 27th, '62; Wake co.

Joseph J. Young, A. Q. M., cm. August 27th, '62; Wake co.
C. B. Justice, A. C. S., cm. August 27th, '62; Wake co.

T. J. Boykin, Surgeon, Virginia.
William Gaither, Surgeon, Burke co.

John E. Moore, Sergeant Major; p. from ranks of Company B.

COMPANY A.

OFFICERS.

Amos B. Cox, Captain, cm. May 17th, '61; Virginia.

A. N. McMillan, Captain, cm. August 10th, '61; Ashe co.; w. at New Bern March 14th, '62.

Samuel P. Wagg, Captain, cm. April 17th, '62; Ashe co.

A. B. Duval, Captain, cm. in '64; Ashe co.

George R. Reeves, 1st Lieut., cm. May 17th, '61; Virginia.

A. B. Duval, 1st Lieut., cm. April 17th, '62; Ashe co.; p.

J. B. Houck, 1st Lieut., cm. July 4th, '63; Ashe co.; w. at Gettysburg and Bristoe Station.

Jesse A. Reeves, 2nd Lieut., cm. May 17th, '61; Virginia.

John C. Plummer, 2nd Lieut., cm. May 17th, '61; Ashe co.

James Porter, 2nd Lieut., cm. August 15th, '61; Ashe co.

Colton Blevins, 2nd Lieut., cm. April 17th, '62; Ashe co.; r. August 5th, '62,

J. B. Houck, 2nd Lieut., cm. April 17th, '62; Ashe co.; p.

NON-COMMISSIONED OFFICERS.

A. N. McMillan, 1st Sergeant, e. May 17th, '61; Ashe co.; p. Captain August 10th, '61; w. at New Bern.

Samuel P. Wagg, 2nd Sergeant, e. May 17th, '61; Ashe co.; p. Captain April 17th, '62.

Samuel Rhea, 3rd Sergeant, e. May 17th, '61; Tennessee; tr. to Cavalry July 20th, '61.

G. A. P. Nye, 4th Sergeant, e. May 17th, '61; Ashe co.; w. July 1st, '63, at Gettysburg.

T. S. Doughton, 5th Sergeant, e. May 17th, '61; Alleghany co.

W. B. Reeves, 1st Corporal, e. May 17th, '61; Ashe co.

F. M. Woody, 2nd Corporal, e. May 17th, '61; Ashe co; dg. March 25th, '62.

James N. Lane, 3rd Corporal, e. May 17th, '61; Wilkes co.; p. Sergeant April 17th, '62; w. at Gettysburg.

M. M. Woody, 4th Corporal, e. May 17th, '61; Ashe co.; k. March 14th, '62, at New Bern.

PRIVATES.

Akers, S. H., e. May 17th, '61; Virginia; d. October 28th, '62, in Virginia.

Ashley, Josiah, e. May 17th, '61; Ashe co.; p. Sergeant July 23d, '62; w. at Gettysburg.

Ashley, J. P., e. March 27th, '62; Ashe co.; p. Sergeant.

Ashley, James, e. March 27th, '62; Ashe co.

Ashley, William, e. March 27th, '63; Ashe co.

Adams, Creed, e. May 17th, '61; Virginia; d. October, '61, at Carolina City.

Baker, Shadrach, e. May 17th, '61; Ashe co.; tr. to Cavalry July 20th, '61.

Baker, Ambrose, e. May 17th, '61; Ashe co.

Baker, James, e. May 17th, '61; Ashe co.; k. July 1st, '63, at Gettysburg.

Baker, J. C., e. May 17th, '61; Virginia; d. of w. received at Gettysburg.

Bear, Lee, e. May 17th, '61; Ashe co.; k. July 1st, '63.

Burgess, Thomas, e. May 17th, '61; Ashe co.; w. and pr. at Gettysburg.
Blackman, Jere., e. March 27th, '62; Ashe co.; tr.
Belvin, Wilborn, e. October 9th, '62; Ashe co.; w. October 14th, '63, at Bristoe Station.
Blevins, Colton, e. May 17th, '61; Ashe co.; p. 2d Lieut. April 17th, '62; r. August 5th, '62.
Blevins, Felix, e. May 17th, '61; Ashe co.; w. at Gettysburg.
Blevins, Horton, e. May 17th, '61; Ashe co.; w. at Gettysburg.
Blevins, Albert, e. March 27th, '62; Ashe co.; w. at Gettysburg.
Blevins, Bartlett, e. March 27th, '62; Ashe co.; pr. at Bristoe Station.
Blevins, Tobias, e. March 27th, '62; Ashe co.; p. Lieut. in Cavalry.
Blevins, Stephen, e. October 9th, '62; Ashe co.; k. July 3d, '63, at Gettysburg.
Blevins, Levi, e. March 15th, '63; Ashe co.
Bear, Alfred, e. October 9th, '62; Ashe co.; k. May 1st, '63, at Weldon by railroad accident.
Burkett, Christian, e. May 17th, '61; Ashe co.; c. July, '63, at Gettysburg.
Benken, Edward, e. March 27th, '62; Ashe co; k. July 3rd, '63, at Gettysburg.
Benken, James, e. March 27th, '62; Ashe co.; d. May 15th, '62, at Bristoe Station
Benken, Jackson, e. March 27th, '62; Ashe co.; d. May 8th, '63, at Kinston.
Ballou, Hugh, e. March 27th, '62; Ashe co.; k. July 1st, '63, at Gettysburg.
Ballou, Lee, e. March 15th, '63; Ashe co.; w. at Gettysburg.
Billings, John, e. July 17th, '61; Wilkes co.
Bowen, George, e. May 17th, '61; Ashe co.
Bear, Randolph, e. May 17th, '61; Ashe co.
Bear, Joseph, e. March 15th, '63; Ashe co.
Bear, James, e. March 15th, '63; Ashe co.
Bowen, Eli, e. May 17th, '61; Ashe co.; d. May, '62, in Ashe co.
Burke, W. E., e. May 17th, '61; Randolph co.; dg.
Baldwin, Joseph, e. October 9th, '62; Ashe co.; w. July 3rd, '63, at Gettysburg.
Black, Michael, e. October 9th, '62; Ashe co.; w. July 1st, '63, at Gettysburg.
Ball, Benjamin, e. March 27th, '64; Ashe co.
Bass, Jesse, e. May 6th, '63; Ashe co.
Calloway, A. F., e. April 27th, '64; Ashe co.
Calloway, James M., e. May 17th, '61; Ashe co.; dg. Jan. 1st, '62.
Calloway, Hiram, e. March 27th, '62; Ashe co.; w. at Gettysburg.
Carter, Isaac M., e. May 17th, '61; Virginia; p. Sergeant April 10th, '62, and w. at Gettysburg.
Coster, T. J., e. May 17th, '61; Virginia; d. April, '62.
Clark, A., e. March 15th, '63; Ashe co.; w. July 3d, '63, at Gettysburg.
Campbell, H. H., e. March 27th, '62; Ashe co.; w. at Gettysburg.
Duval, James M., e. May 17th, '61; Ashe co.; p. Sergeant April 17th, '62, and w. at Bristoe Station.
Duval, James, e. March 27th, '62; Ashe co.
Duval, Maston, e. March 27th, '62; Ashe co.; w. July 1st, '63, at Gettysburg.
Duval, John, e. March 27th, '62; Ashe co.; d. June 24th, '62, at Goldsboro.
Demry, Uriah, e. May 17th, '61; Yadkin co.; w. and pr. at Gettysburg.
Denny, Walter, e. October 9th, '62; Ashe co.
Duval, A. B., e. May 17th, '61; Ashe co.; p. Sergeant August 10th, '61, and 1st Lt. April 17th, '62.
Davis, George W., e. May 17th, '61; Virginia; dg. April 25th, '62.
Davis, James, e. May 17th, '61; Virginia.
Davis, William, e. May 17th, '61; Virginia; w. at Gettysburg.
Daughton, Joseph, e. May 17th, '61; Alleghany co.; p. Corporal February, '62; pr. at Falling Waters.
Dickson, James, e. Sept. 1st, '61; Ashe co.; d. June 12th, '62, at Petersburg.

Dillard, V., e. Oct. 9th, '62; Ashe co.
Edmonston, Abram, e. May 17th, '61; Burke co.; tr. to Com. E April 17th, '62.
Etheridge, John, e. March 27th, '62; Ashe co.; w. at Rolls' Mill and Gettysburg.
Foster, Joel E., e. May 17th, '61; Burke co.
Foster, E. F., e. May 17th, '61; Wilkes co.; w. at Gettysburg.
Fisher, G. R., e. May 17th, '61; Wilkes co.; pr. at Gettysburg.
Fields, John G., e. May 17th, '61; Alleghany co.; w. at Gettysburg and pr.
Furman, Noah, e. March 27th, '62; Ashe co.
Furman, Calloway, e. March 27th, '62; k. October 14th, '63, at Bristoe Station.
Gentry, Joseph R., e. May 17th, '61; Ashe co.; w. and pr. at Gettysburg.
Greer, John F., e. May 17th, '61; Ashe co.
Grimsley, Lowry, e. May 17th, '61; Ashe co.; w. at Gettysburg.
Grimsby, Drury, e. May 17th, '61; Ashe co.; tr.
Gentry, Nichols, e. May 17th, '61; Ashe co.; k. Oct. 14th, '63, at Bristoe Station.
Grimsley, Hiram, e. May 17th, '61; Ashe co.; d. Sept. 10th, '61, at Raleigh.
Gambrell, J. F., e. May 17th, '61; Ashe co.; d. March, '62, at Goldsboro.
Graybeal, John, e. March 27th, '62; Ashe co.
Hawthorn, A. C., e. March 27th, '62; Ashe co.; w. at Gettysburg.
Harless, George, e. May 17th, '61; Ashe co.; w. at Bristoe Station.
Harless, Shadrach, e. May 17th, '61; Ashe co.; d. of w. received near Richmond.
Hash, John E., e. May 17th, '61; Ashe co.; w. at Bristoe Station.
Hash, James, e. May 17th, '61; Ashe co.
Houck, J. C., e. May 17th, '61; Ashe co.; p. Sergeant April 17th, '63.
Houck, Washington, e. March 27th, '62; Ashe co.; w. July 3rd, '63, at Gettysburg.
Houck, J. W., e. March 27th, '62; Ashe co.; w. at Gettysburg.
Hurley, N. P., e. May 17th, '61; Virginia.
Houck, J. B., e. May 17th, '61; Ashe co.; p. Sergeant July 20th, '61; 2d Lieut. April 7th, '62; 1st Lieut. July 4th, '63; w. at Gettysburg and Bristoe Station.
Houdy, John, e. March 27th, '62; Ashe co.; d. May 15th, '62, at Kinston.
Hurley, Marshall, e. March 15th, '63; Ashe co.; k. July 1st, '63, at Gettysburg.
Ingram, John C., e. May 17th, '61; Wilkes co.
Jones, Lewis S., e. May 17th, '61; Ashe co.; p. 2d Sergeant April 17th, '62.
Ketchum, J. W., e. May 17th, '61; Ashe co.
Ketchum, George, e. Sept. 1st, '61; Ashe co.; d. Nov., '61, in Carolina City.
Lewis, John, e. March 27th, '62; Ashe co.
Little, George, e. May 17th, '61; Ashe co.
Logan, Naaman, e. May 17th, '61; Ashe co.; w. at Gettysburg.
Long, James W., e. May 17th, '61; Alleghany co.
Landreth, Isaac W., e. May 17th, '61; Alleghany co.
Miller, Joseph, e. May 17th, '61; Ashe co.; d. March 14th, '62, at New Bern.
McMillan, William, e. May 17th, '61; Ashe co.
Miller, Harrison, e. May 17th, '61; Ashe co.; pr. July 3d, '63, at Gettysburg.
Miller, Jonathan, e. October 9th, '62; Ashe co.; k. July 3d, '63, at Gettysburg.
Miller, J. H., e. October 9th, '62; Ashe co.; d. March 14th, '63, in Hospital.
Miller, S., e. October 9th, '62; Ashe co.
Miller, Isham, e. October 9th, '62; Ashe co.
Mitchell, T., e. March 15th, '63; Ashe co.; w. at Gettysburg.
Mink, G. R., e. October 9th, '62; Ashe co.
Mahab, Marshall, e. May 17th, '61; Ashe co.
Osborne, Jasper, e. May 17th, '61; Ashe co.
Osborne, Lee, e. May 17th, '61; Ashe co.; p. 1st Lieutenant in Cavalry September 1st, '63.
Osborne, Zach., e. May 17th, '61; Ashe co.
Osborne, Granville, e. May 17th, '61; Ashe co.; w. at Gettysburg and Richmond.
Osborne, C. A., e. Sept. 1st, '61; Alleghany co.; d. Feb. 20th, '62, at Petersburg.

Pennington, William, e. March 27th, '62; Ashe co.
Plummer, Samuel, e. Sept. 1st, '61; Ashe co.; dg. August 10th, '62.
Price, John, e. May 17th, '61; Ashe co.; k. July 1st, '62, at Malvern Hill.
Plummer, J. R., e. May 17th, '61; Ashe co.; tr. to Cavalry July 20th, '62.
Porter, James, e. May 17th, '61; Ashe co.; p. 2d Lieutenant August 15th, '61.
Porter, William R., e. May 17th, '61; Ashe co.; dg. May 10th, '62.
Porter, William, e. March 15th, '63; Ashe co.
Plummer, J. M. C., e. May 17th, '61; Virginia; w. at Gettysburg.
Phipps, Benjamin, e. May 17th, '61; Ashe co.
Phipps, Nathan, e. May 17th, '61; Ashe co.
Pendergrass, ——, e. November 24th, '63; Ashe co.
Pendergrass, Joshua, e. March 27th, '62; Ashe co.; w. at Gettysburg.
Penny, Calvin, e. October 9th, '62; Ashe co.
Plummer, M. M., e. May 17th, '61; Ashe co.; d. July 16th, '63, of w. received at Gettysburg.
Perkins, Ambrose, e. March 27th, '62; Ashe co.; w. July 1st, '63, at Gettysburg.
Perkins, W. H., e. March 27th, '62; Ashe co.; dg. July 20th, '63.
Perkins, Riley, e. March 27th '62; Ashe co.; m. at Gettysburg.
Perkins, Timothy, e. March 27th, '62; Ashe co.; k. July 1st, '63, at Gettysburg.
Reedy, C. N., e. March 27th, '62; Ashe co.; w. at Gettysburg.
Roten, Ashley, e. May 17th, '61; Ashe co.
Roten, Jonathan, e. Sept. 1st, '61; Ashe co.; d. Dec., '61.
Roten, Solomon, e. October 9th, '62; Ashe co.; k. July 3d, '63, at Gettysburg.
Richardson, J., e. Sept. 1st, '61; Ashe co.; dg.
Rutherford, Thomas, e. May 17th, '61; Ashe co.
Reeves, A. M., e. May 17th, '61; Ashe co.; p. 1st Lieutenant in Cavalry September 11th, '63; w. at Gettysburg.
Reeves, William, e. May 17th, '61; Virginia; m. at Gettysburg.
Robertson, James H., e. May 17th, '61; Virginia; w. at Bristoe Station.
Richardson, C., e. May 17th, '61; Virginia; d. April, '62.
Richardson, Elbert, e. May 17th, '61; Virginia; d. at Carolina City.
Reid, Jackson, e. May 17th, '61; Wilkes co.
Riley, James A., e. March 27th, '64; Ashe co.
Stamper, John, e. May 17th, '61; Ashe co.; w. at Gettysburg.
Stamper, Ira, e. May 17th, '61; Ashe co.; w. at Gettysburg.
Stamper, M. D., e. May 17th, '61; Ashe co.
Senter, J. C., e. May 17th, '61; Ashe co.; k. at Bristoe Station.
Stringer, Cameron, e. May 17th, '61; Ashe co.
Sheets, Joseph, e. May 17th, '61; Ashe co.; d. at New Bern.
Sheets, Troy, e. May 17th, '61; Ashe co.; p. 4th Corporal; pr. at Bristoe Station.
Sheets, David, e. March 15th, '63; Ashe co.
Sheets, James, e. March 15th, '63; Ashe co.
Sheets, A., e. October 9th, '62; Ashe co.; w. at Gettysburg.
Smith, Elijah F., e. May 17th, '61; Virginia; p. 2d Corporal April 17th, '62; w. at Gettysburg.
Sullen, Ambrose, e. March 27th, '62; Ashe co.; d. May 20th, '63.
Smith, A. B., e. May 17th, '61; Ashe co.; dg. August, '62.
Sanders, Alexander, e. May 17th, '61; Ashe co.
Senter, John, e. Sept. 1st, '61; Ashe co.; dg. August 5th, '62.
Stedham, Samuel, e. May 17th, '61; Ashe co.; d.
Stedham, Lewis, e. May 17th, '61; Ashe co.; w. July 1st, '63, at Gettysburg.
Stoud, Abram, e. February 2d, '64; Ashe co.
Sullen, Joseph, e. March 27th, '62; Ashe co.; dg. in '63, for disability.
Testament, T. M., e. March 27th, '62; Ashe co.; k. July 1st, '63, at Gettysburg.
Turner, J. H., e. March 27th, '62; Ashe co.; w. at Gettysburg.

Taylor, Cline, e. October 9th, '62; Ashe co.; w. July 1st, '63, at Gettysburg.
Taylor, William H., e. March 27th, '62; Ashe co.; p. Corporal.
Taylor, William, e. May 17th, '61; Ashe co.; k. March 14th, '62, at New Bern.
Taylor, Calloway, e. May 17th, '61; Ashe co.; k. July 1st, '63, at Gettysburg.
Taylor, J. R., e. May 17th, '61; Ashe co.; d. of w. received at Roles' Mill.
Taylor, W. H., e. May 17th, '61; Alleghany co.; w. July 1st, '63, at Gettysburg.
Treadway, W. W., e. May 17th, '61; Ashe co.; p. Sergeant; w. at Gettysburg.
Vanover, John, e. May 17th, '61; Ashe co.
Wagoner, Samuel, e. May 17th, '61; Ashe co.; p. 3d Corporal April 17th, '62; w. at Gettysburg.
Wagoner, H. D., e. May 17th, '61; Ashe co.; k. April 20th, '62.
Weiss, R. M., e. May 17th, '61; Virginia; d. April 22d, '62, at Goldsboro.
Wyatt, Reid, e. May 17th, '61; Ashe co.; p. Lieut. in Cavalry; dg.
Wagg, A. W., e. March 27th, '62; Ashe co.
Walton, P. M., e. March 27th, '62; Ashe co.
Walker, H. Z., e. March 27th, '62; Ashe co.
Young, E. S., e. May 17th, '61; Ashe co.; pr. July 3d, '63, at Gettysburg.
Yates, S. S., e. May 17th, '61; Wilkes co.; w. at Bristoe Station.

COMPANY B.

OFFICERS.

J. J. C. Steele, Captain, cm. June 5th, '61; Union co.
Thos. J. Cureton. Captain, Union co. p. from 1st Lieut.

Wm. Wilson, 1st Lieut., cm. June 5th, '61; Union co.
T. J. Cureton, 1st Lieut., Union co.
John W. Richardson, 1st Lieut., cm. Jan. 5th, '63; k. at Gettysburg.
N. B. Edwards, 1st Lieut., Union co.
T. J. Cureton, 2d Lieut., cm. June 5th, '61; Union co.
W. W. Richardson, 2d Lieut., cm. June 5th, '61; Union co ; p. from ranks.
Wm. M. Estridge, 2d Lieut., cm. July 2d, '63; p. from ranks; w. July 1st, '63.
A. B. Edwards, 2d Lieut., cm. in '63; w. July 1st, '63.

NON-COMMISSIONED OFFICERS.

J. H. Neely, 1st Sergeant, e. June 5th, '61; Union co.; d. April 22d, '62.
Thos. J. Cureton, 2d Sergeant, e. June 5th, '61; Union co.; p.
Calvin Dickerson, 3d Sergeant, e. June 5th, '61; Union co.
Geo. Richards, 4th Sergeant, e. June 5th, '61; Union co.
Robert F. Jennings, 5th Sergeant, e. June 5th, '61; Union co.; d. Sept. 1st, '61.
Samuel H. Walkup, 1st Corporal, e. June 5th, '61; Union co.; p. 1st Sergeant in January, '63; w. twice.
Burwell B. Griffin, 2d Corporal, e. June 5th, '61; Union co.; d. Sept., '62.
E. C. McCain, 3d Corporal, e. June 5th, '61; Union co., dg. Aug. 20th, '61.
H. B. W. McNeely, 4th Corporal, e. June 5th, '61; Union co.; d. April 11th, '62.
Brazel B. Johnson, Musician, e. June 5th, '61; Union co.

PRIVATES.

Alexander, George N., e. June 5th, '61; Union co.
Ackroid, William, e. June 5th, '61; Union co.; dg.
Austin, William H., e. June 5th, '61; Union co.
Austin, Green D., e. June 5th, '61; Union co.; w. July 3d, '63, at Gettysburg.
Baker, W. A., e. June 5th, '61; Union co.; dg. September 10th, '61.
Baker, Thomas A., e. June 5th, '61; Union co.; dg.
Baker, William T., e. December 10th, '64; Wilkes co.
Broom, Thomas J., e. June 5th, '61; Union co.; pr. July 3d, '63, at Gettysburg.
Broom, William J., e. June 5th, '61; Union co.
Broom, A. L., e. June 5th, '61; Union co.; pr. July 3d, '63, at Gettysburg.
Broom, Jackson, e. June 5th, '61; Union co.; tr. to Com. B, 43d Regiment.
Boyle, McF., e. June 17th, '61; Union co.
Boyle, J. T. C., e. June 17th, '61; Union co.; dg. April 23d, '62.
Blackaney, Ruse, e. June 5th, '61; Union co.
Burke, James, e. September 28th, '62; Catawba co.
Byram, Robert, e. February 14th, '63; Union co.
Belk, Noah A., e. June 5th, '61; Union co.; d. November 5th, '61.
Britz, Edward A., e. January 12th, '63; Forsyth co.
Cook, James M., e. June 5th, '61; Union co.
Cook, George W., e. June 5th, '61; Union co.; k. July 1st, '62, at Malvern Hill.
Cooper, John, e. June 5th, '61; Union co.; dg.
Costles, Moses D., e. June 5th, '61; Union co.
Clanton, William E., e. Sept, 21st, '62; Wilkes co.
Clark, John B., e. August 1st, '61; Union co.; dg. Oct. 19th, '62, at Petersburg.
Church, C. A., e. Sept. 21st, '62; Wilkes co.
Church, H. M., e. Sept. 21st, '62; Wilkes co.
Church, J. L., e. Sept. 21st, '62; Wilkes co.; pr. Oct. 27th, '64.
Church, J. R., e. Sept. 21st, '62; Wilkes co.
Cosley, P. P., e. June 5th, '61; Union co.; dg. October, '61.
Chancey, E. P., e. August 17th, '61; Union co.
Cox, William F., e. Sept. 21st, '62; Wilkes co.
Dearson, David, e. June 5th, '61; Union co.; dg. Sept. 5th, '61.
Dulin, James, e. October 10th, '62; Mecklenburg co.; d. May 1st, '63, at Greenville.
Dulin, John, e. October 10th, '62; Mecklenburg co.
Duse, Charles F., e. June 5th, '61; Union co.; k. July 1st, '62, at Malvern Hill.
Davis, Robert A., e. June 5th, '61; Union co.; dg. August 20th, '61.
Easom, Alfred, e. August 17th, '61; Union co.
Easom, Asa, e. June 5th, '61; Union co.; pr. July 3d, '63, at Gettysburg.
Estep, Hardy, e. Sept. 21st, '62; Wilkes co.; k. July 1st, '63, at Gettysburg.
Estep, Doctor, e. Sept. 21st, '62; Wilkes co.
Eller, James F., e. Sept. 21st, '62; Wilkes co.
Estridge, Wm. M., e. June 5th, '61; Union co.; p. 2d Lieut. July 2d, '63.
Festerman, Henry T., e. June 5th, '61; Union co.
Fincher, James A., e. June 5th, '61; Union co.; d. July 10th, '63, at Gettysburg.
Freizeland, W. R., e. March 28th, '63; Union co.
Givens, Samuel A., e. June 5th, '61; Union co.
Givens, Hugh, e. June 5th, '61; Union co.; d. April 14th, '62, at Kinston.
Griffin, John, e. June 5th, '61; Union co.; k. July 1st, '63, at Gettysburg.
Griffin, James L., e. August 17th, '61; Union co.; dg. April 22nd, '62.
Glenn, William H., e. June 5th, '61; Union co.
Glenn, Robert S., e. December 29th, '62; Union co.
Glenn, George L., e. March 28th, '63; Union co.
Gay, Isaac E., e. June 5th, '61; Union co.; dg. December 19th, '61.

Gordon, Thomas C., e. December 24th, '62; Union co.; pr. October 27th, '64.
Hilton, Leroy R., e. June 5th, '61; Union co.
Hayes, Shelton, e. June 5th, '61; Union co.; d. July 1st, '63, of w. received at Gettysburg.
Helms, Charles F., e. August 8th, '61; Union co.
Helms, Ezekiel, e. August 8th, '61; Union co.; d. December 19th, '62.
Helms, Noah E., e. January 21st, '62; Union co.; tr. from Com. B, 43rd Regiment; d. December 9th, '62.
Huffstrickler, D. S., e. June 5th, '61; Cleveland co.
Huffstrickler, John A., e. June 5th, '61; Cleveland co.
Hamby, Benjamin W., e. September 21st, '62; Wilkes co.
Honeycut, F. L., e. August 1st, '61; Union co.; k. July 1st, '63, at Gettysburg.
Honeycut, Wilborn, e. September 8th, '61; Union co.
Hinson, S. M., e. June 5th, '61; Union co.; d. August 10th, '63.
Hill, M. T., e. August 17th, '61; Union co.; dg. April 25th, '62.
Inman, W. A., e. June 5th, '61; Union co.
Johnson, John A., e. September 28th, '62; Wilkes co.; dg. August 19th, '63.
Johnson, Rufus, e. December 10th, '64; Wilkes co.
Joines, Hampton, e. September 21st, '62; Wilkes co.
Knight, Burwell J., e. June 5th, '61; Union co.; pr. July 3rd, '63, at Gettysburg.
Kirkland, Daniel, e. June 5th, '61; Union co.; d. August 10th, '62.
Lanney, William B., e. June 5th, '61; Union co.; d. July 10th, '63, of w. received at Gettysburg.
Lanney, Samuel L., e. June 5th, '61; Union co.
Lanney, John, e. June 5th, '61; Union co.; pr. July 3d, '63, at Gettysburg.
Lanney, John B., e. June 5th, '61; Union co.; k. July 3d, '63, at Gettysburg.
Lanney, John, e. August 1st, '61; Union co.; dg. Sept. 5th, '62.
Long, Adam, e. August 17th, '61; Union co.; d. April 15th, '62.
Lowry, Wm. F., e. June 5th, '61; Union co.
Lowe, J. M., e. Sept. 21st, '62; Wilkes co.
McCain, John J., e. June 5th, '61; Union co.; d. August 26th, '62.
McCain, John S., e. June 5th, '61; Union co.; pr. October 27th, '64.
McCorkle, Wm. H., e. June 5th, '61; Union co.; pr. October 14th, '63.
McCorkle, John W., e. August 1st, '61; Union co.
McGirst, Isham S., e. June 5th, '61; Union co.
McWhorter, John, e. June 5th, '61; Union co.; k. July 1st, '63, at Gettysburg.
Moseley, Wm. C., e. August 17th, '61; Union co.; d. April 25th, '63.
McManus, Columbus, e. June 5th, '61; Union co.; pr. October 27th, '64.
McManus, Henry J., e. June 5th, '61; Union co.
McManus, George W., e. June 5th, '61; Union co.
McRory, W. H., e. August 17th, '61; Union co.; dg. Nov., '64.
McRory, Thomas M., e. Nov. 1st, '61; Union co.; k. March 14th, '62, at New Bern.
Mulis, Marshall J., e. June 5th, '61; Union co.; p. Corporal and pr. Oct. 27th, '64.
Mulis, Solomon S., e. August 17th' '61; Union co.; drowned March 14th, '63, in Briar Creek.
Matton, Isaac, e. June 5th, '61; Union co.
Moore, John E., e. June 5th, '61; Union co.; p. Sergeant Major in 26th Regiment July 1st, '62.
Miller, Thomas F. M., e. June 5th, '61; Union co.; dg. September 5th, '62.
McNeely, James M., e. June 5th, '61; Union co.; dg. August 10th, '61.
Mills, Tighlman D., e. September 21st, '62; Wilkes co.
Mangum, H. J., e. May 3d, '64; Stokes co.; pr. Oct. 27th, '64.
Mangum, T. B., e. March 14th, '63; Wilkes co.
Meekle, Joseph, e. March 29th, '64; Wilkes co.
Norwood, W. E., e. Oct. 24th, '64; Stokes co.

Nichols, J. B., e. September 21st, '62; Wilkes co.
Osborne, Alexander, e. June 5th, '61; Union co.
Osborne, Milton C., e. June 5th, '61; Union co.
Phillips, William, e. June 5th, '61; Union co.
Phillips, Benjamin F., e. June 5th, '61; Union co.
Pardue, William, e. June 5th, '61; Union co.
Pique, Amos, e. June 5th, '61; Union co.
Preslar, Darlin, e. February 14th, '63; Union co.
Richardson, B. J., e. June 5th, '61; Union co.
Richardson, S. L., e. June 5th, '61; Union co.
Richardson, S. D., e. August 1st, '61; Union co.; w. July 1st, '63, at Gettysburg.
Richardson, J. M., e. Dec. 24th, '61; Union co.; pr. July 3d, '63, at Gettysburg.
Richardson, G. W., e. June 5th, '61; Union co.; dg. October, '61.
Richardson, W. W., e. June 5th, '61; Union co.; p. 2d Lieutenant January 5th. '63; k. July 1st, '63, at Gettysburg.
Richards, Thomas, e. December 4th, '61; Union co.
Robinson, James H., e. June 5th, '61; Union co.
Robinson, Robert, e. June 5th, '61; Union co.
Rollins, John D., e. June 5th, '61; Union co.; p. Color Bearer; w. July 1st, '63, at Gettysburg; d. October 26th, '64.
Rogers, James, e. June 5th., '61; Union co.; d. October 20th, '63, of w. received at Bristoe Station.
Rogers, Anderson, e. June 5th, '61; Union co.; pr. Oct. 27th, '64.
Rogers, Robert, e. Oct. 18th, '61; Union co.; d. Aug. 21st, '63, of w. received July 1st, '63, at Gettysburg.
Rope, James P., e. June 5th, '61; Union co.; d. April 1st, '62, at Bristoe Station.
Ray, John J., e. Sept 21st, '62; Wilkes co.
Russion, W. P., e. Dec. 24th, '62; Union co.; d. April 19th, '63, at Petersburg.
Staton, Neill B., e. June 5th, '61; Union co.
Starnes, Wm., e. June 5th, '61; Union co.
Starnes, Wm., e. June 5th, '61; Union co.
Starnes, Nathaniel, e. June 5th, '61; Union co.
Starnes, Thos., e. June 5th, '61; Union co.; d. July 21st, '63, of w. received at Gettysburg.
Starnes, M. L., e. Feb. 14th, '63; Union co.
Starnes, J. F. L., e. Sept. 6th, '62; Union co.; d. Dec. 12th, '61.
Starnes, W. P. R, e. June 5th, '61; Union co ; dg.
Starnes, Alexander, e. June 5th, '61; Union co.; dg. Oct. 10th, '61.
Smith, Wm. E., e. June 5th, '61; Union co.
Sikes, Marcus, e. June 5th, '61; Union co.
Simpson, J. R., e. June 5th, '61; Union co.
Simms, William, e. June 5th, '61; Union co.; dg, September 5th, '63.
Secrest, John A., e. August 1st, '61; Union co.; d. April 19th, '62, at Goldsboro.
Secrest, Leroy S., e. August 1st, '61; Union co.
Thomas, Hampton H., e. June 5th, '61; Union co.; w. July 3d, '63, at Gettysburg.
Thompson, William H., e. June 5th, '61; Union co.; d. April 25th, '62.
Tucker, D., e. September 28th, '62; Union co.; k. October 14th, '63, at Bristoe Station.
Tarlton, C. L., e. June 5th, '61; Union co.; dg. Oct. 10th, '61.
Underwood, William M., e. June 5th, '61; Union co.; dg. September 5th, '62.
Williams, A. L., e. June 5th, '61; Union co.; dg. August 20th, '61.
Walkup, Henry L., e. June 5th, '61; Union co.; w. July 1st, '63, at Gettysburg.
Walkup, Israel P., e. December 29th, '62; Union co.
Walkup, J. C., e. Oct. 24th, '64; Stokes co.
Welsh, John, e. Dec. 10th, '64; Stokes co.

Whitaker, Atlas, e. May 3d, '64; Stokes co.
Whitaker, Thomas C., e. June 5th, '61; Union co.
White, B. R., e. December 29th, '62; Union co.
Wyan, S. W., e. September 28th, '61; Catawba co.; d. December 27th, '64.
Waters, Philip, e. June 5th, '61; Union co.; dg. September 5th, '62.
Woody, S. B., e. September 21st, '62; Wilkes co.
Woody, Theophilus, e. September 22d, '62; Wilkes co.
Wyatt, Andrew, e. September 21st, '62; Wilkes co.
Wyatt, Adam, e. September 21st, '62; Wilkes co.; k. July 1st, '63, at Gettysburg.
Wyatt, J. W., e. September 21st, '62; Wilkes co.; d. December, '62, at Petersburg.
Wyatt, Jonas, e. September 21st, '62; Wilkes co.
Wounger, William C., e. September 21st, '62; Wilkes co.

COMPANY C.

OFFICERS.

Abner R. Carmichael, Captain, cm. June 12th, '61; Wilkes co.; p. Major and k. at New Bern March 14th, '62.
Alexander H. Horton, Captain, cm. August 30th, '61; Wilkes co.
J. A. Jarrett, Captain, cm. in '64; Wilkes co.

Calvin Church, 1st Lieut., cm. June 12th, '61; Wilkes co.; r. Sept. 30th, '61.
William W. Hampton, 1st Lieut., cm. October 5th, '61; Wilkes co.
William Porter, 1st Lieut., cm. April 21st, '62; Wilkes co.; pr. in '64.
Phinehas Horton, 2d Lieut., cm. June 12th, '61; Wilkes co.
W. W. Hampton, 2d Lieut., cm. June 12th, '61; Wilkes co ; p.
Thomas L. Furgerson, 2d Lieut., cm. August 30th, '61; Wilkes co.
Alexander H. Horton, 2d Lieut., cm. August 30th, '61; Wilkes co.; p. Captain.
James A. Hague, 2d Lieut., cm. October 5th, '61; Wilkes co.
J. A. Jarrett, 2d Lieutenant, cm. April 21st, '62; Yadkin co.; p. from ranks; p. Captain.
Rufus D. Horton, 2d Lieut., cm. December 12th, '62; Wilkes co.
John M. Harris, 2d Lieut., cm. April 21st, '62; Wilkes co.

NON-COMMISSIONED OFFICERS.

Thos. L. Furgerson, 1st Sergeant, e. June 12th, '61; Wilkes co.; p. 2d Lieut. April 21st, '62.
Jesse F. Triplett, 2d Sergeant, e. June 12th, '61; Wilkes co.; p. 1st Sergeant; k. July 3d, '63, at Gettysburg.
John M. Harris, 3d Sergeant, e. June 12th, '61; Wilkes co.; p. 2d Lieut. April 21st, '62.
Samuel L. Pryor, 4th Sergeant, e. June 12th, '61; Wilkes co.; p. April 21st, '62.
John F. Walters, 5th Sergeant, e. June 12th, '61; Wilkes co.; d. November, '61, at Carolina City.
E. P. Coffey, 1st Corporal, e. June 12th, '61; Wilkes co.; d. December 19th, '61, at Morehead City.
James A. Hague, 2d Corporal, e. June 12th, '61; Wilkes co.; p. 2d Lieutenant Sept. 27th, '61.

David Hall, 3d Corporal, e. June 12th, '61; Wilkes co.; p. Sergeant April 21st, '62; w. July 3d, '63, at Gettysburg.

W. H. Parker, 4th Corporal, e. June 12th, '61; Wilkes co.; d. November 24th, '61, at Carolina City.

PRIVATES.

Abshire, Levi, e. June 12th, '61; Wilkes co.; pr. in '64.

Adams, Jacob, e. June 12th, '61; Wilkes co.; d. February 28th, '62, at New Bern.

Adams, J. A., e. September 30th, '62; Wilkes co.

Adams, F. M., e. September 30th, '62; Wilkes co.; w. July 1st, '63, at Gettysburg; pr. in '64.

Adams, W. H., e. September 30th, '62; Wilkes co.

Alley, F. M., e. December 11th, '62; Wilkes co.; July 1st, '63, at Gettysburg.

Anton, S., e. December 11th, '62; Mecklenburg co.

Anton, Benjamin F., e. December 11th, '62; Mecklenburg co.; w. July 1st, '63, at Gettysburg.

Ball, James, C., e. June 12th, '61; Wilkes co.; w. July 1st, '63, at Gettysburg.

Bangus, Reuben, e. June 12th, '61; Wilkes co.; dg. August 1st, '61, for disability.

Bell, John, R., e. June 12th, '61; Wilkes co.; w. July 1st, '63, at Gettysburg.

Bell, Evans, e. June 12th, '61; Wilkes co.; w. at Malvern Hill; dg. August 21st, '62.

Blankenship, William, e. June 12th, '61; d. April 12th, '62, at Kinston.

Brown, M. C., e. June 12th, '61; Wilkes co.

Bullis, Benjamin F., e. June 12th, '61; Wilkes co.; p. 4th Corporal April 21st, '62.

Bullis, Wesley, e. March 20th, '63; Wilkes co.

Bullis, S., e. June 12th, '61; Wilkes co.; w. and c. July 3d, '63, at Gettysburg; d. in prison.

Byrd, Josiah, e. June 12th, '61; Wilkes co.; dg. August 9th, '61, for disability.

Bumgarner, A., e. March 21st, '63; Wilkes co.; k. July 1st, '63, at Gettysburg.

Barlow, T. L., e. September 21st, '62; Wilkes co.

Barlow, H. H. e. September 21st, '62; Wilkes co.; w. July 1st, '63, at Gettysburg.

Church, A. W., e. June 12th, '61; Wilkes co.; w. at Malvern Hill.

Church, Leander, e. June 12th, '61; Wilkes co.; c. and d. a prisoner at New Bern.

Church, Harrison, e. June 12th, '61; Wilkes co.; dg. August 9th, '61.

Chapel, F. T., e. June 12th, '61; Wilkes co.; d. August 20th, '63, of w. received at Gettysburg.

Combs, Wesley, e. June 12th, '61; Wilkes co.

Combes, Wesley, e. June 12th, '61; Wilkes co.; pr. in '64.

Combes, C., e. September 1st, '61; Wilkes co.

Cox, E., e. March 20th, '63; Wilkes co.; w. July 3d. '63, at Getteysburg.

Cranfill, John W., e. June 12th, '61: Wilkes co.; k. July 1st, '63, at Gettysburg.

Curtis, James, e. June 12th, '61; Wilkes co.; k. July 3d, '63, at Gettysburg.

Curtis, Wm., e. June 12th, '61; Wilkes co.; p. Ordnance Sergeant; w. at Malvern Hill July 1st, '62.

Dancy, Calvin, e. June 12th, '61; Wilkes co.; dg. August 25th, '62.

Dancy, David E., e. June 12th, '61; Wilkes co.

Davis, Thomas, e. June 12th, '61; Wilkes co.; w. at Malvern Hill July 1st, '63.

Davis, A. M., e. September 21st, '62; Wilkes co.; w. July 3d, '63, at Gettysburg; d. in '63.

Dickson, John, e. October 17th, '62; Yakin co.; w. July 3d, '63, at Gettysburg; pr. in '64.

Dockery, James, e. June 12th, '61; Wilkes co.; d. Nov. 27th, '61, at Carolina City.

Dudley, James, e. June 12th, '61; Wilkes co.; w. July 1st, '63, at Gettysburg.

Edmonston, James, e. June 12th, '61; Wilkes co.; d. Oct. 31st, '63, of w. received at Gettysburg and Bristoe Station.

Edmonston, William, e. June 12th, '61; Wilkes co.; pr. in '64.

Edmonston, John, e. June 12th, '61; Wilkes co.; pr. in '64.

Edmonston, John, e. June 12th, '61; Wilkes co.; w. at New Bern.

Edmonston, Abram, e. May 17th, '61; Ashe co.; w. July 1st, '63, at Gettysburg; pr. in '64.

Ellen, James M., e. June 12th, '61; Wilkes co.; pr. in '64.

Ellen, Jacob M., e. June 12th, '61; Wilkes co.

Ellen, Leandor, e. March 5th, '62; Wilkes co.; tr. from Company K, 53d Regiment; pr. in '64.

Ferguson, Jesse T., e. June 12th, '61; Wilkes co.; p. Sergeant August 30th, '61; Commissary Sergeant March 1st, '62; pr. in '64.

Foster, John R., e. June 12th, '61; Wilkes co.; w. July 1st, '63, at Gettysburg.

Foster, A., e. September 1st, '63; Wilkes co.

Foster, J. H., e. September 21st, '62; Wilkes co.

Faircloth, Philip, e. November 1st, '63; Wilkes co.; pr. in '64.

Faircloth, L., e. November 4th, '63; Orange co.

German, H., e. September 14th, '63; Wilkes co.

Gilbert, Joshua, e. June 12th, '61; Wilkes co.; d. at Petersburg.

Gilbert, Charles, e. November 15th, '61; Wilkes co.; p. Color Bearer; k. October 14th, '63, at Bristoe Station.

Griffith, Barnett, e. June 12th, '61; Wilkes co.; dg. August 9th, '61, for disability.

Griffith, Thomas, e. June 12th, '61; Wilkes co.; tr. to Com. K, 53d Regiment.

Hall, Franklin, e. June 12th, '61; Wilkes co.

Hall, Henry, e. June 12th, '61; Wilkes co.

Hall, James E., e, June 12th, '61; Wilkes co.

Hall, John, e. June 12th, '61; Wilkes co.; dg. December 6th, '61, for disability.

Hall, Riley R., e. June 12th, '61; Wilkes co.

Hall, P., e. September 21st, '62; Wilkes co.

Hall, Samuel P., e. June 12th, '61; Wilkes co.; w. July 3d, '63, at Gettysburg.

Hall, T. D., e. June 12th, '61; Wilkes co.; p. Sergeant.

Hall, J. G., e. September 21st, '63; Wilkes co.; d. December, '62, in Hospital.

Hamby, Triplet, e. June 12th, '61; Wilkes co.; w. July 1st, '63, at Gettysburg.

Handy, W. H., e. June 12th, '61; Wilkes co.

Higgins, Esley, e. June 12th, '61; Wilkes co.; p. 1st Sergeant July 1st, '63.

Higgins, Hiram, e. June 12th, '61; Wilkes co.

Higgins, W. W., e. June 12th, '61; Wilkes co.

Hinson, John, e. June 12th, '61; Stanly co.; pr. in '64.

Harris, Elisha, e. June 12th, '61; Wilkes co.

Hollar, Henry, e. June 12th, '61; Wilkes co.; dg. August 9th, '61, for disability.

Horton, Rufus D., e. December 12th, '62; Wilkes co.; p. 2d Lieutenant December 12th, '62; w. July 3d, '63, at Gettysburg.

Horton, A. H. I., e. June 12th, '61; Wilkes co.; p. 2d Lieutenant August 6th, '61.

Holder, W. H., e. June 12th, '61; Wilkes co.; w. July 1st, '63, at Malvern Hill.

Harless, James, e. November 13th, '63; Wilkes co.; pr. in '64.

Harless, W. H., e. June 12th, '61; Wilkes co.; pr. in '64.

Harding, A., e. December 11th, '62; Wilkes co.

Jennings, J. D., e. September 21st, '62; Wilkes co.

Jarratt, J. A., e. October 8th, '61; Yadkin co.; p. 2d Lieut. April 21st, '62.

Johnson, James, e. June 12th, '61; Wilkes co.; w. July 1st, '63, at Gettysburg.

Joins, Eli, e. June 12th, '61; Wilkes co.; w. July 1st, '63, at Gettysburg.

Keetur, F. M., e. October 17th, '62; Wilkes co.; k. July 1st, '63, at Gettysburg.

Kemp, G. E., e. December 11th, '62; Wilkes co.; k. July 1st, '63, at Gettysburg.

Keeley, M., e. December 11th, '62; Wilkes co.; k. March 14th, '63, at New Bern.

Lewis, Joshua, e. June 12th, '61; Wilkes co.; k. October 14th, '63, at Bristoe Station.

Lewis, Joseph, e. June 12th, '61; Wilkes co.; pr. in '64.

Lewis, James, e. September 21st, '62; Wilkes co.; w. July 1st, '63, at Gettysburg.

Lewis, B. F., e. September 21st, '62; Wilkes co.; k. July 1st, '63, at Gettysburg.

Lewis, T., e. September 21st, '62; Wilkes co.; d. December, '63, in Hospital.

Lee, Charles, e. June 12th, '61; Wilkes co.

Laws, B., e. September 21st, '62; Wilkes co.

Moss, H., e. December 11th, '62; Gaston co.; w. July 3d, '63, at Gettysburg; pr. in '64.

Martin, J. W., e. December 11th, '62; Mecklenburg co.; d. June 18th, '63, in Hospital.

McDaniel, F. H., e. June 12th, '61; Wilkes co.; w. July 1st, '63, at Gettysburg.

Milton, James G., e. June 12th, '61; Wilkes co.; p. Color Bearer and w. July 1st, '63, at Gettysburg.

Millsaps, Josiah, e. June 12th, '61; Wilkes co.

Minton, M. C., e. June 12th, '61; Wilkes co.

Minton, Thomas, e. June 12th, '61; Wilkes co.; w. July 1st, '63, at Gettysburg.

Minton, Thomas C., e. June 12th, '61; Wilkes co.; tr. to Company D, 26th Regiment, April 21st, '62.

Moore, Wesley, e. June 12th, '61; Wilkes co.; dg. for disability.

Mullis, R. S., e. June 12th, '61; Wilkes co.; d. Nov. 3d, '63, at Carolina City.

Mullis, Thomas, e. June 12th, '61; Wilkes co.; d. Dec. 15th, '63, at Richmond.

Monroe, Alexander, e. December 11th, '62; Wilkes co.; pr. in '64.

Maddel, M., e. September 21st, '62; Wilkes co.

Nance, Benjamin A., e. June 12th, '61; Wilkes co.; w. at Malvern Hill.

Nance, Wm., e. June 12th, '61; Wilkes co.; w. July 1st, '63, at Gettysburg; p. Corporal in '64.

Nance, James Y., e. June 12th, '61; Wilkes co.

Nichols, Anderson, e. June 12th, '61; Wilkes co.; k. July 3d, '63, at Gettysburg; pr. in '64.

Nichols, David, e. June 12th, '61; Wilkes co.

Nichols, James, e. June 12th, '61; Wilkes co.; d. June 12th, '62, at Carolina City.

Owens, George W., e. June 12th, '61; Wilkes co.

Owens, W. H., e. June 12th, '61; Wilkes co.

Parsons, Alfred, e. June 12th, '61; Wilkes co.; w. July 1st, '63, at Gettysburg; pr. in '64.

Parsons, Peyton, e. June 12th, '61; Wilkes co.; w. July 1st, '63, at Gettysburg.

Parsons, W. F., e. March 5th, '62; Wilkes co.; tr. from 53d Regiment; w. July, '63, at Gettysburg.

Parsons, W. N., e. March 20th, '63; Wilkes co.

Pierce, J. H. C., e. June 12th, '61; Wilkes co.; dg. August, '62.

Porter, William, e. June 12th, '61; Wilkes co.; p. 1st Lieutenant April 21st, '62.

Pryor, S. L., e. June 12th, '61; Wilkes co.; p. 1st Sergeant in '64.

Pryor, C. M., e. September 21st, '62; Wilkes co.; tr. to Com. K, 53d Regiment.

Reeves, John, e. June 12th, '61; Wilkes co.; w. July 1st, '63, at Gettysburg.

Reese, S., e. September 21st, '62; Wilkes co.; tr. to Com. K, 53d Regiment.

Rupard, Caleb, e. December 11th, '62; Wilkes co.; w. July 1st, '63, at Gettysburg; pr. in '64.

Robertson, W. W., e. March 20th, '63; Wilkes co.

Rhodes, W. W., e. September 21st, '62; Wilkes co.; pr. in '64.

Rangus, Henry, e. October 19th, '61; Wilkes co.

Ray, John, e. June 12th, '61; Wilkes co.; w. at Malvern Hill; pr. in '64.

Simmons, C., e. June 12th, '61; Wilkes co.; pr. in '64.

Simmons, Thomas, e. June 12th, '61; Wilkes co.; k. July 1st, '63, at Gettysburg.

Stanly, Robert A., e. June 12th, '61; Wilkes co.
Sneed, W. H., e. June 12th, '61; Wilkes co.; dg. August 9th, '61, for disability.
Souther, Joseph, e. September 21st, '62; Wilkes co.
Souther, Jesse, e. September 21st, '62; Wilkes co.; w. July 3d, '63, at Gettysburg.
Souther, John, e. September 21st, '62; Wilkes co.
Souther, Henry, e. September 21st, '62; Wilkes co.
Southern, Joshua, e. June 12th, '61; Wilkes co.; w. July 1st, '62, at Malvern Hill; w. July 3d, '63, at Gettysburg; pr. in '64.
Shell, David, e. June 12th, '61; Wilkes co.; dg. August 9th, '62.
Shumatt, Thomas, e. September 21st, '62; Wilkes co.
Tedler, Joel H., e. June 12th, '61; Wilkes co.
Thorpe, David, e. June 12th, '61; Wilkes co.; dg. August 9th, '62.
Triplett, A. L., e. September 21st, '62; Wilkes co.; w. July 1st, '63, at Gettysburg; p. Sergeant in '64.
Triplett, E., e. September 21st, '62; Wilkes co.
Triplett, W., e. September 21st, '62; Wilkes co.; d. December, '63.
Triplett, H. S., e. September 21st, '62; Wilkes co.; w. July 1st, '63, at Gettysburg.
Triplett, J. W., e. September 21st, '62; Wilkes co.; d. November 18th, 63, of w. received October 14th, '63, at Bristoe Station.
Triplett, Moses, e. March 5th, '62; Wilkes co.
Watts, William H., e. November 3d, '64; Wilkes co.
Watts, T. J., e. September 21st, '62; Wilkes co.; w. July 1st, '63, at Gettysburg.
Wright, Joseph W., e. June 12th, '61; Wilkes co.; p. Corporal in '61.
Williams, James, e. June 12th, '61; Wilkes co.
Williams, H., e. September 21st, '62; Wilkes co.
Williams, W., e. September 1st, '63; Wilkes co.
Walker, W., e. March 20th, '63; Wilkes co.
Welch, James, e. October 17th, '62; Wilkes co.
Welch, T. B., e. March 1st, '63; Wilkes co.; pr. in '64.
Welch, W. P., e. June 12th, '61; Wilkes co.; w. July 1st, '63, at Gettysburg; p. Sergeant in '64.
Welch, George, e. June 12th, '61; Wilkes co.; p. 3d Corporal August 21st, '62.
Welch, Bennett H., e. June 12th, '61; Wilkes co.; p. Corporal in '64.
Wells, W. D., e. June 12th, '61; Wilkes co.
Wooten, John, e. June 12th, '61; Wilkes co.; dg. December 6th, '61.
Yates, Alfred, e. June 12th, '61; Wilkes co.
Yates, J. F., e. June 12th, '61; Wilkes co.
Yates, J. D., e. September 21st, '62; Wilkes co.
Yoster, P. H., e. June 12th, '61; Wilkes co.; w. July 3d, '63, at Gettysburg.

COMPANY D.

OFFICERS.

Oscar R. Rand, Captain, cm. May 22d, '61; Wake co.; pr. at New Bern March 14th, '62.
T. J. Adams, Captain, cm. April 12th, '62; Wake co.; p. Major.

James B. Jordan, 1st Lieut., cm. May 22d, '61; Wake co.
H. S. Whitaker, 1st Lieut., cm. ———; Wake co.
Gaston H. Broughton. 1st Lieut., cm. April 28th, '62; Wake co.
James S. Adams, 2d Lieut., cm. May 22d, '61; Wake co.
James W. Vincent, 2d Lieut., cm. May 22d, '61; Wake co.
James G. M. Jones, 2d Lieut., cm. April 29th, '62; Wake co.
Wm. N. Snelling, 2d Lieut., cm. April 29th, '62; Wake co.

NON-COMMISSIONED OFFICERS.

Daniel G. Beckwith, 1st Sergeant, e. May 22d, '61; Wake co.; pr. at New Bern and d.
Joseph J. Young, 2d Sergeant, Wake co.; p. A. Q. M. March 20th, '62.
J. G. Banks, 3d Sergeant, e. May 22d, '61; w. at New Bern.
Gaston H. Broughton, 4th Sergeant, e. May 22d, '61; Wake co.; p. 1st Lieut.
James G. M. Jones, 5th Sergeant, e. May 22d, '61; Wake co.; p. 2d Lieut.
Wm. C. Booker, 1st Corporal, e. May 22d, '61; Wake co.; w. at Gettysburg.
Wm. H. Boothe, 2d Corporal, e. May 22d, '61; Wake co.; d. of w. received at Gettysburg.
James Q. Adams, 3d Corporal, e. May 22d, '61; Wake co.; d. March 10th, '62, at New Bern.
Lucius J. Gulley, 4th Corporal, e. May 22d, '61; Wake co.; dg.

PRIVATES.

Atkins, Saul, e. August 19th, '61; Wake co.; pr. in '64.
Austin, Simeon, e. May 29th, '61; Wake co.
Austin, Sidney, e. March 8th, '63; Wake co.
Austin, Samuel A., e. May 29th, '61; Wake co.; p. Sergeant.
Adams, David C., e. June 16th, '62; Wake co.; w. at Bristoe Station.
Booker, William E., e. May 19th, '62; Wake co.; w. at Gettysburg; p. Sergeant in '64.
Bryan, N. F., e. May 29th, '61; Wake co.; k. July 3d, '63, at Gettysburg.
Burt, Jesse G., e. May 29th, '61; Wake co.; w. at Bristoe Station.
Burt, Willie P., e. May 29th, '61; Wake co.; p. Corporal June 12th, '62; p. 1st Sergeant in '64.
Burt, William P., e. May 29th, '61; Wake co.
Burt, R. Quent, e. May 29th, '61; Wake co.; p. Corporal.
Burt, Joseph H., e. August 19th, '61; Wake co.; k. July 1st, '63, at Gettysburg.
Broughton, Wm., e. June 10th, '61; Wake co.; d. October 14th, '61, at Carolina City.
Bonden, Moses, e. June 10th, '61; Wake co.; dg.
Beckwith, Henderson, e. May 29th, '61; Wake co.; dg. for disability.
Baker, Alexander, e. June 10th, '61; Wake co.

Baker, Wesley, e. June 10th, '61; Wake co.; w. at Gettysburg.
Baker, Joseph, e. June 10th, '61; Wake co.; w. at Gettysburg.
Baker, Jackson, e. August 19th, '61; Wake co.
Baker, Orren, e. October 3d, '62; Wake co ; d. May, '63, at Goldsboro.
Bowling, G. W., e. September 22d, '62; Wake co.; w.
Bell, William W., e. October 1st, '62; Wake co.; pr. in '63.
Brooks, George, e. September 22d, '62; Wake co.
Bynum, H. M., e. January 20th, '63; Wake co.
Barber, Andrew, e. January 20th, '63; Wake co.; d. March, '63, at Goldsboro.
Booker, A. B., e. October 16th, '63; Wake co.; pr. in '64.
Booker, W. J., e. May 29th, '61; Wake co.; p. Sergeant.
Clements, John, e. October 22d, '64; Wake co.
Chesson, James H., e. October 22d, '64; Wake co.
Carroll, Thos. C., e. May 29th, '61; Wake co.; w. at Gettysburg; pr. in '64.
Carroll, Wiley, e. May 29th, '61; Wake co.
Champion, William, e. May 29th, '61; Wake co.; k. October 14th, '63, at Bristoe Station.
Crawford, Dallas, e. August 19th, '61; Wake co.; w. Oct. 14th, '63, at Bristoe Station.
Camps, John, e. September 22d, '62; Wake co.; w. July 1st, '63, at Gettysburg.
Camps, Abram, e. September 22d, '62; Wake co.
Chamblee, John, e. October 1st, '62; Wake co.
Chisholm, Alexander, e. October 18th, '62; Wake co.
Dancey, Joseph, e. September 22d, '62; Wake co.; w. at Gettysburg.
Ennis, Joseph, e. August 19th, '62; Wake co.
Ellis, Anderson, e. September 22d, '62; Wake co.; pr. in '63.
Everby, Martin, e. May 29th, '61; Wake co.
Ennis, J. W., e. October 22d, '64; Wake co.
Ford, A. J., e. May 29th, '64; Wake co.
Fuquay, John, e. May 29th, '61; Wake co.; dg. for disability.
Fuquay, Stephen, e. June 10th, '62; Wake co.; w. at Gettysburg.
Ford, Anderson, e. October 1st, '62; Wake co.
Goodwin, N. G., e. May 29th, '61; Wake co.
Goodwin, S. W., e. May 29th, '61; Wake co.; w. at Gettysburg.
Gower, E. P., e. May 29th, '61; Wake co.
Gilbert, H. Z., e. May 29th, '61; Wake co.; k. July 3d, '63, at Gettysburg.
Gilmore, James, e. January 20th, '63; Wake co.; w. at Bristoe Station.
Gill, David S., e. January 20th, '63; Wake co.
Gill, B. Thomas, e. September 22d, '63; Wake co.
Griffice, George, e. May 29th, '61; Wake co.; pr. and d.
Gower, William, e. May 29th, '61; Wake co.; pr. and d.
Gower, Perrin, e. August 19th, '61; Wake co.; dg.
Geringer, P. H., e. October 16th, '63; Wake co.; pr. in '64.
Hunter, Reuben, e. May 29th, '61; Wake co.
Hunter, Thomas, e. May 29th, '61; Wake co,
Handy, Y. B., e. September 22d, '62; Wake co.
Handy, Noah, e. September 22d, '62; Wake co.
Hall, A. B., e. September 22d, '62; Wake co.
Hamilton, Jacob, e. September 22d, '62; Wake co.; d. in North Carolina.
Holt, E, J., e. May 29th, '61; Wake co.
Hawkins, Reuben, e. Sept. 22d, '62; Wake co.
Hamilton, Mills G., e. May 29th, '61; Wake co.; dg.
Hamilton, Wesley, e. May 29th, '61; Wake co.; w. at Gettysburg; pr. in '64.
Holleman, T. M., e. May 29th, '61; Wake co.; pr. in '64.
Holleman, J. Q., e. Sept. 11th, '61; Wake co.; d. Dec., '61, at New Bern.

Holleman, Silas G., e. Sept. 11th, '61; Wake co.; dg.
Honeycut, J. P., e. May 29th, '61; Wake co.
Henry, W. P., e. May 29th, '61; Wake co.
Harvel, John, e. May 29th, '61; Wake co.; d. March, '62, at New Bern.
Hicks, H. C., e. Aug. 19th, '61; Wake co.; tr. to the 31st Regiment.
Jones, Walter A., e. May 29th, '61; Wake co.
Jones, R. K., e. October 16th, '63; Wake co.
Jones, Sidney L., e. Aug. 19th, '61; Wake co.
Jones, Alexander, e. Jan. 20th, '63; Wake co.; d. in '64.
Jones, Fanning, e. May 29th, '61; Wake co.; d. in N. C.
Jones, John L., e. June 8th, '63; Wake co.; w. at Bristoe Station.
Johnson, H. L., e. Aug. 19th, '61; Wake co.
Jennings, Reuben, e. Sept. 22d, '62.
Jenkins, John, e. Sept. 22d, '62; Wake co.; d. in Hospital at Petersburg.
Jenkins, David, e. Sept. 22d, '62; Wake co.; w. at Gettysburg.
Kelly, Joseph, e. May 29th, '61; Wake co.; d. March 14th, '62, at New Bern.
Kelly, George, e. May 29th, '61; Wake co.
King, Wm. H., e. March 16th, '63; Wake co.
Langston, H. W., e. May 29th, '61; Wake co.
Langston, W. W., e. May 26th, '61; Wake co.
Langston, F. M., e. May 26th, '61; Wake co.
Langston, J. P., e. March 8th, '63; Wake co.
Lee, Gallington W., e. September 22d, '63; Wake co.
McLenmore, Duncan, e. May 29th, '61; Wake co.; dg. for disability.
Mills, Wiley J., e. May 29th, '61; Wake co.
Morris, George, e. January 20th, '63; Wake co; pr. in '63.
McDonald, William, e. October 1st, '62; Wake co.; w. at Gettysburg.
Mangum, L. D., e February 10th, '62; Wake co.
Marlow, Albert, e. September 22d, '62; Wake co.
Maynard, Quent., e. June 20th, '62; Wake co.
Mayler, Jacob, e. October 16th, '63; Wake co.
Maynard, Sidney, e. September 22d, '62; Wake co.; d. at Hanover Junction, Va.
Menton, Jesse, e. September 22d, '62; Wake co.
Matthew, Etheldridge, e. August 19th, '61; Wake co.
Nordan, John B., e. June 10th, '61; Wake co.
Nordan, Nathan, e. June 10th, '61; Wake co.
Nordan, William R., e. June 10th, '61; Wake co.; d. May 16th, '62, in N. C.
Norris, Needham, e. May 29th, '61; Wake co; dg.
Norris, George, e. August 19th, '61; Wake co.; d. in Carolina City.
Norris, William B., e. September 11th, '61; Wake co.; p. Corporal; pr. in '61.
Olive, Asa, e. May 29th, '61; Wake co.; d. October, '62, in N. C.
Oliver, Thomas, e. September 22d, '62; Wake co.; k. at Gettysburg.
Partin, John H., e. May 29th, '61; Wake co.
Partin, G. W., e. May 29th, '61; Wake co.; w.
Pope, Jackson, e. May 29th, '61; Wake co.; k. at New Bern.
Pearson, Solomon, e. May 29th, '61; Wake co.
Prince, John, e. November 16th, '62; Wake co.
Pollard, S., e. May 29th, '61; Wake co.
Pollard, B. A., e. May 19th, '62; Wake co.
Powell, Alexander, e. May 29th, '61; Wake co.; dg. for disability.
Rhodes, James F., e. May 29th, '61; Wake co; d. of w. at Goldsboro.
Ray, Henderson, e. August 19th, '61; Wake co.
Rogers, Wm. C., e. February 16th, '61; Wake co.
Smith, Addison, e. May 29th, '61; Wake co.; d. at Carolina City.

Smith, Wm. K., e. May 29th, '61; Wake co.; k. October 14th, '62, at Bristoe Station.

Snelling, Wm. N., e. June 10th, '61; Wake co.; p. 1st Sergeant May, '62; p. Lieut. July 3d, '63.

Stephens, James A., e. May 29th, '61; Wake co.

Stephens, James L., e. May 29th, '61; Wake co.; w.

Smith, John, e. September 22d, '62; Wake co.; dg. for disability.

Seagraves, Wesley, e. May 29th, '61; Wake co.

Seagraves, Paschal, e. May 29th, '61; Wake co.; c. and d.

Seagraves, Wm., e. May 29th, '61; Wake co.; c. and d.

Stevens, T., e. May 29th, '61; Wake co.; dg.

Stephenson, James, e. March 20th, '62; Wake co.; d. November 12th, '62, at Petersburg.

Shepherd, John, e. September 22d, '62; Wake co.

Stephens, J. A., e. May 29th, '64; Wake co.; pr. in '64.

Temple, Alfred H., e. May 29th, '61; Wake co.; w. twice.

Temple, Isham, e. May 29th, '61; Wake co ; dg.

Turner, John F., e. May 29th, '61; Wake co.; d. at Carolina City.

Thompson, Benjamin, e. May 29th, '61; Wake co.; d. at Carolina City.

Trensan, Rufus, e. Sept. 22d, '62; Wake co.

Treadway, Eli, e. Sept. 22d, '62; Wake co.

Tighlman, James, e. November 4th, '63; Wake co.; k.

Twinson, Joseph, e. May 29th, '61; Wake co.; dg.

Utley, Wm. F., e. May 29th, '61; Wake co.; p. Sergeant May 22d, '63.

Utley, Britton S., e. May 29th, '61; Wake co.; p. Sergeant May 22d, '63.

Utley, James W. T., e. May 29th, '61; Wake co.; dg. for disability.

Wood, Cornelius, e. May 29th, '61; Wake co.

Wheeler, Joseph, e. May 19th, '62; Wake co.

Whitaker, F. A., e. Aug. 19th, '62; Wake co.; dg.

Whitaker, H. S., e. Aug. 19th, '62; Wake co.; p. 1st Lieut.

Williams, O. S., e. May 29th, '61; Wake co.; dg.

Williams, Sidney, e. August 19th, '61; Wake co.; c. and d.

Wilson, Joseph, e. August 19th, '61; Wake co.; dg.

Womble, John, e. May 29th, '61; Wake co.; c. and d. at New Bern.

Wilson, Lewis, e. May 29th, '61; Wake co.; dg. for disability.

Wyatt, Sidney, e. Sept. 22d, '62; Wake co.

Wyatt, William, e. Sept 22d, '62; Wake co.

Wyatt, Jordan, e. Sept. 22d, '62; Wake co.

Wyatt, Jesse, e. Sept. 22d, '62; Wake co.

Wyatt, John, e. Sept. 22d, '62; Wake co.

Wyatt, Nathan, e. Sept. 22d, '62; Wake co.

Wingler, John, e. Sept. 22d, '62; Wake co.; k. Oct. 14th, '63, at Bristoe Station.

Wingler, Elijah, e. Sept. 22d; '62; Wake co.

Wingler, R., e. Sept. 22d, '62; Wake co.; k. July 1st, '63, at Gettysburg.

Williams, A. J., e. September 22d, '62; Wake co.

Wyatt, W. D., e. October 16th, '63; Wake co.

Williams, Augustus, e. September 22d, '62; Wake co.

Williams, Marshall, e. October 1st, '62; Wake co.

Williams, Kizey, e. October 1st, '62; Wake co.

Wallace, James K., e. June 20th, '62; Wake co.

Young, Josiah D., e May 29th, '61; Wake co.; p. Corporal.

Young, Samuel H., e. May 29th, '61; Wake co.

Young, Simon Y., e. May 29th, '61; Wake co.

COMPANY E.

OFFICERS.

W. S. Webster, Captain, cm. May 28th, '61; Chatham co.
S. W. Brewer, Captain, cm. ——; Chatham co.; pr. in '64.

W. J. Headen, 1st Lieut., cm. May 28th, '61; Chatham co.
S. W. Brewer, 1st Lieut., Chatham co.; p.
W. J. Lambert, 1st Lieut., Chatham co.
R. G. Dunlap, 2d Lieut., cm. May 28th, '61; Chatham co.
S. W. Brewer, 2d Lieut., cm. May 28th, '61; Chatham co.; p.
John B. Emerson, 2d Lieut., cm. April 12th, '62; Chatham co.
W. J. Lambert, 2d Lieut., cm. April 12th, '62; Chatham co.; pr. in '64.
Owen A. Hannor, 2nd Lieut., cm. in October, '63; p. from ranks.
E. H. McManus, 2d Lieut., cm. in '64; Chatham co.; pr. in '64.

NON-COMMISSIONED OFFICERS.

Sidney J. Tally, 1st Sergeant, e. May 28th, '61; Chatham co.; p. 2d Lieut. in 44th Regiment.
W. H. Merritt, 2d Sergeant, e. May 18th, '61; Chatham co.; w. at Gettysburg.
James M. Brooks, 3d Sergeant, e. May 28th, '61; Chatham co.; p. 1st Sergeant; pr. in '64.
John W. Calder, 4th Sergeant, e. May 28th, '61; Chatham co.; w. June, '62, and at Gettysburg.
James H. McMath, 5th Sergeant, e. May 28th, '61; Chatham co.; w. at Malvern Hill and Gettysburg.
J. W. McDaniel, 1st Corporal, e. May 28th, '61; Chatham co.; w. at Gettysburg.
John B. Emerson, 2d Corporal, e. May 28th, '61; Chatham co.; p. 2d Lieut. April 12th, '62.
Thos. H. Phillips, 3d Corporal, e. May 28th, '61; Chatham co.; dg. Nov. 11th, '61.
John W. Dowd, 4th Corporal, e. May 28th, '61; Chatham co.; w. at Gettysburg.

PRIVATES.

Adcock, J. B., e. in March, '61; Chatham co.; d. October 15th, '63, of w. received at Bristoe Station.
Adcock, John W., e. in March, '61; Chatham co.; w. at Gettysburg.
Andrews, James, e. May 28th, '61; Chatham co.; d. September 10th, '63, at Morehead City.
Andrews, James, e. May 28th, '61; Chatham co.; w. at Gettysburg.
Andrews, Joseph J., e. May 28th, '61; Chatham co.; k. at Gettysburg.
Adcock, J. W., e. March 8th, '62; Chatham co.
Blaylock, J. C., e. October 28th, '64; Chatham co.
Bowen, John, e. October 28th, '64; Chatham co.
Brewer, Isham W., e. February 10th, '64; Chatham co.
Beal, Benjamin M., e. May 28th, '61; Chatham co.; w. at Bristoe Station.
Beal, J. W., e. May 28th, '61; Chatham co.
Beaver, Daniel, e. May 28th, '61; Chatham co.; w. at Gettysburg; pr. in '64.
Borough, William H., e. May 28th, '61; Chatham co.
Brady, Christenburg, e. May 28th, '61; Chatham co.
Brafford, Atlas, e. in March, '63; Chatham co.

Brafford, James, e. May 28th, '61; Chatham co.; dg. July 16th, '62.
Brafford, Robert, e. in September, '62; Chatham co.
Brantley, William T., e. May 28th, '61; Chatham co.; d. July, '62.
Bray, Henry H., e. May 28th, '61; Chatham co.; d. April, '62.
Brewer, Eli C., e. May 28th, '62; Chatham co.
Brewer, Elijah, e. in March, '63; Chatham co.; w. October 14th, '63, at Bristoe Station.
Brewer, Jere. G., e. May 28th, '61; Chatham co.
Brewer, Henry, e. in March, '62; Chatham co.; k. July 1st, '62, at Malvern Hill.
Brewer, Nathan, e. in March, '62; Chatham co.; w. July 1st, '63, at Gettysburg.
Bright, J. W., e. May 28th, '61; Chatham co.; d. June, '63, at Lynchburg.
Brooks, Josiah Y., e. May 28th, '61; Chatham co.; dg. September 1st, '63.
Brooks, John A., e. May 28th, '61; Chatham co.; w. at Gettysburg.
Burkes, G. Brooks, e. in April, '63; Chatham co.; tr. from Company H; w. at Gettysburg.
Burkes, William H., e. May 23rd, '61; Chatham co.; dg.
Bowen, G. G., e. May, '61; Chatham co.; p. Corporal; pr. in '64.
Burroughs, W. H., e. August 20th, '61; Chatham co.
Burke, Thomas B., e. September 1st, '62; Chatham co.
Bright, John B., e. October 28th, '64; Chatham co.
Brown, Adam, e. September 22d, '62; Chatham co.
Calder, J. W., e. April 18th, '64; Chatham co.
Cheek, Wesley G., e. October 28th, '64; Chatham co.
Campbell, M. H., e. May 28th, '61; Chatham co.
Cabe, John, e. September, '62; Chatham co.; pr. in '64.
Carter, Daniel M., e. May 28th, '61; Chatham co.; d. at Winchester July 12th, '63, of w. received at Gettysburg.
Carter, John H., e. March, '62; Chatham co.; w. July 1st, '63, at Gettysburg.
Carter, W. David, e. March 28th, '62; Chatham co.; w. at Barrington's Ferry March, '63, and at Gettysburg.
Caveness, William W., e. May, '62; Moore co.; w. July 1st, '63, at Gettysburg.
Clariday, Joseph C., e. March, '63; Chatham co.; w. July 1st, '63, at Gettysburg.
Cheek, Middleton, e. January, '63; Chatham co.; tr. from Com. G, 44th Regiment; w. at Get ysburg.
Cheek, Randal, e. May 28th, '61; Chatham co.; w. at Gettysburg.
Cheek, Robert, e. May 28th, '61; Chatham co.; w. at Gettysburg.
Cheek, William M., e. May 28th, '61; Chatham co.
Crutchfield, J. H., e. May 28th, '61; Chatham co.; dg.
Crutchfield, William, e. May 28th, '61; Chatham co.
Davis, Elwood, e. May 28th, '61; Chatham co.; w. October 10th, '61.
Darsett, James D., e. May 28th, '61; Chatham co.; p. Corporal April, '62; w. at Gettysburg.
Darsett, James M., e. March, '62; Chatham co.; d. November, '62, at Petersburg.
Darsett, S. J., e. March, '62; Chatham co.; w. at Gettysburg.
Darsett, S. S., e. May 28th, '61; Chatham co.; d. May 18th, '62.
Dixon, Richard, W., e. October 22d, '64; Chatham co.
Dowd, John W., e. May 28th, '61; Chatham co.
Dowd, W. J., e. April 18th, '64; Chatham co.
Doughty, Nicholas, e. October 28th, '64; Chatham co.
Ellington, G. E., e. December 8th, '64; Chatham co.
Emerson, J. P., e. March 8th, '62; Chatham co.
Elphins, Joseph, e. October 28th, '61; Chatham co.
Edwards, J. N., e. January, '63; Chatham co.; p. Corporal; w. at Gettysburg.
Edwards, William W., e. May 28th, '61; Chatham co.; p. Corporal; w. at Gettysburg.

Ellington, Farrington, e. October, '61; d. March, '62.
Ellington, Wm. P., e. Oct., '61; Chatham co.; k. at Gettysburg.
Ellis, Joseph C., e. July, '61; Chatham co.
Ellis, James N., e. May 28th, '61; Chatham co.; p. Sergeant April, '62.
Ellis, John W., e. May 28th, '61; Chatham co.; p. Sergeant April, '62.
Ellis, Laban, e. May 28th, '62; Chatham co.; w. at Gettysburg.
Ellis, Wm. B., e. February, '63; Chatham co.; d. July, '63, at Richmond.
Ellis, Wm. M., e. May 28th, '61; Chatham co.; dg. in '61.
Evans, O. H., e. May 28th, '61; Chatham co.; k. June 25th, '62.
Fields, Green, e. May 28th, '61; Chatham co.; w. at Gettysburg.
Fields, James N., e. May 28th, '61; Chatham co.; w. at Gettysburg.
Fitts, George H., e. May 28th, '61; Chatham co.; p. Sergeant.
Foster, Alfred H., e. May 28th, '61; Chatham co.
Foster, David, e. May 28th, '61; Chatham co.; k. at Gettysburg.
Foster, Manly, e. March, '62; Chatham co.; w. at Gettysburg.
Foster, Nathaniel, e. May 28th, '61; Chatham co.; w. at Gettysburg.
Fields, James, e. October 29th, '64; Chatham co.
Forrester, H. H., e. May 22d, '61; Chatham co.; d. in '64.
Goldston, J. D., e. July 12th, '64; Chatham co.; w. at Spottsylvania C. H.
Gee, Wm. H., e. March, '62; Chatham co ; w. at Gettysburg and Bristoe Station.
Gilliland, James O., e. March, '63; Chatham co.
Gilliland, Walter W., e. March, '63; Chatham co.; pr. in '64.
Hackney, Joseph D., e. May 28th, '61; Chatham co.; dt. Drum Major.
Hackney, Thomas C., e. May 28th, '61; Chatham co.
Hadley, Newton, e. Oct., '61; Chatham co.
Hanner, Owen A., e. May 28th, '61; Chatham co.; w. at Malvern Hill; p. 2d Lieut. October, '63.
Harper, Joseph S., e. Sept., '62; Chatham co.; k. at Gettysburg.
Harper, Wm. W., e. May 28th, '61; Chatham co.; w. at Gettysburg.
Harris, Noah R., e. May 28th, '61; Chatham co.; dg.
Hart, John, e. May 28th, '61; Chatham co.; k. at Gettysburg.
Hill, James R., e. May 28th, '61; Chatham co.
Hamalay, E. H., e. October, '61; Chatham co.; p. Ord. Sergt. May, '62.
Howard, A. B., e. September, '62; Chatham co.
Hudson, C. David, e. May, '61; Chatham co.; dg. July, '62.
Hackney, Albert, e. October 22d, '64; Chatham co.; pr. at Spottsylvania C. H.
Hall, John, e. October 1st, '64; Chatham co.
Hilliard, Joseph, e. October 8th, '64; Chatham co.
Hicks, William, e. October 28th, '64; Chatham co.
Hackney, W. D., e. October 10th, '64; Chatham co.
Jones, J. E., e. September 22d, '64; Chatham co.
Jenkins, Richard L., e. June 5th, '61; Chatham co.; w. October 14th, '63, at Bristoe Station.
Johnson, Edward, e. May 28th, '61; Chatham co.; dg. September, '62.
Johnson, Isaac N., e. September, '62; Chatham co.; d. January, '63, at Petersburg.
Johnson, James, e. May, '62; Chatham co.; w. at Gettysburg and Bristoe Station.
Jones, John E., e. September, '62; Chatham co.
Jones, John J., e. May 28th, '61; Chatham co.; w. October 14th, '63, at Bristoe Station.
Jones, R. A., e. May 28th, '61; Chatham co.
Jones, William, e. May 28th, '61; Chatham co.; w. at Gettysburg.
Kidd, Neal, e. May 28th, '61; Chatham co.; p. Sergeant.
Lambert, John J., e. May 28th, '61; Chatham co.; w. at Gettysburg.
Lambert, William, e. July, '61; Chatham co.; p. 2nd Lieut. April, '62.

McManns, Edwin H., e. August, '61; Chatham co.; w. at Gettysburg.
Mansfield, Jefferson B., e. May 28th, '61; Chatham co.; p. Color Sergeant and w. at Gettysburg.
Martindale, James. e. May 28th, '61; Chatham co.; w. at Gettysburg and Bristoe Station; p. Corporal; pr. in '64.
Mobly, John, e. May 28th, '61; Chatham co.; k. at Gettysburg.
Moody, Charles, e. Aug., '62; Chatham co.; w. at Gettysburg.
Moody, James, e. May 28th, '61; Chatham co.; w. at Bristoe Station.
Moody, Joseph, e. September, '62; Chatham co.; tr. to Com. H, 46th Regiment.
Marshall, William, e. October 28th, '64; Chatham co.
Myrick, M. Ensley, e. June 15th, '61; Randolph co.; dg. for w. received by accident.
McMath, J. H., e. July 2d, '61; Chatham co.
Nall, William B., e. March 16th, '64; Chatham co.
Nall, Erwin, e. September, '62; Chatham co.
Nall, Martin L., e. May 28th, '61; Chatham co.; w. at Gettysburg; pr. in '64.
Needham, Thomas, e. June 16th. '61; Randolph co.; pr. in '64.
Needham, William, e. June 15th, '61; Randolph co.; w. at Gettysburg.
Norwood, John S., e. June 13th, '61; Chatham co.; w. at Gettysburg; pr. in '64.
Page, Everett, e. May 28th, '61; Chatham co.; k. at Gettysburg.
Page, N. L., e. May 28th, '61; Chatham co.; dg. February 1st, '62.
Perry, Henry H., e. July 1st, '61; Chatham co.; k. at Gettysburg.
Petty, Jesse W., e. July 1st, '61; Chatham co.; w. at Gettysburg and Bristoe Station.
Petty, George D., e. May, '61; Chatham co.
Phillips, George W., e. May, '61; Chatham co.; w. in North Carolina; pr. in '64.
Phillips, John, e. August, '61; Chatham co.; w. at Gettysburg.
Phillips, John Hosea, e. May, '61; Chatham co.; k. at Gettysburg; p. Corporal in '62.
Phillips, Richard, e. May, '61; Chatham co.; w. at Gettysburg.
Phillips, Thomas, e. May, '61; Chatham co.; k. at Gettysburg.
Phillips, Willis R., e. May, '61; Chatham co.
Powers, John W., e. August, '62; Chatham co.; w. at Gettysburg.
Ray, John H., e. May 28th, '61; Chatham co.; k. June 28th, '62.
Rivers, Daniel E., e. May 28th, '61; Chatham co.; dg. May, '62.
Rivers, Joseph E., e. May 28th, '61; Chatham co.; dg. August, '62.
Rivers, William, e. May 28th, '61; Chatham co.; dg. July, '62.
Rogers, Henry M., e. July 5th, '61; Chatham co.; k. at Gettysburg.
Russell, John W., e. May 28th, '61; Chatham co.; k. at Gettysburg.
Smith, Green B., e. May, '62; Chatham co.; k. at Gettysburg.
Smith, John L., e. October, '61; Chatham co.
Smith, L. B., e. May 28th, '61; Chatham co.; w. March 14th, '62, at New Bern.
Smith, Noah R., e. May 28th, '61; Chatham co.; w. at Gettysburg.
Smith, Orren, e. December, '61; Chatham co.; d. December, '63.
Smith, S. A., e. May 28th, '61; Chatham co.; dg.
Smith, Thomas H., e. May 28th, '62; Chatham co.
Smith, W. M., e. May 28th, '61; Chatham co.; d. April, '62, at Kinston.
Smith, W. R., e. May, '62; Chatham co.; d. July 10th, '63, at Winchester, of w. received at Gettysburg.
Sanders, N. R., e. May, '62; Chatham co.
Stanly, Atlas, e. March, '62; Chatham co.; w. at Gettysburg.
Shields, Allen R., e. May, '62; Chatham co.; d. July 30th, '63, of w.
Smith, Robert M., e. May, '61; Chatham co.; w. at Gettysburg.
Seagle, L. S., e. May, '61; Chatham co.; p. Sergeant.
Scott, Franklin J., e. October 28th, '64; Chatham co.; w. in '64.

Seagal, J. C., e. May 22d, '61; Chatham co.
Thomas, Ambrose, e. October 28th, '64; Chatham co.
Thomas, Gabriel, e. October 28th, '64; Chatham co.
Tally, Berry W., e. May, '61; Chatham co.; d. May, '63, at Halifax.
Tally, Robert D., e. May, '62; Chatham co.
Teague, L. T., e. May, '62; Chatham co.; w. at Gettysburg.
Temple, James A., e. May 18th, '61; Chatham co.; k. May, '63, at Halifax.
Thomas, Daniel B., e. May 28th, '61; Chatham co.; w. at Gettysburg.
Thomas, John, e. May 28th, '61; Chatham co.
Thomas, Nathaniel, e. May 28th, '61; Chatham co.
Thomas, Nathaniel, e. October 22d, '64; Chatham co.
Tally, J. Ruffin, e. May 28th, '61; Chatham co.; d. July 8th, '63, of w. received at Gettysburg.
Tally, Milton V., e. May 28th, '61; Chatham co.; tr. to 44th Regiment.
Troy, M. M., e. Aug. 28th, '61; Chatham co.; dg.
Vestal, A. B., e. March, '63; Chatham co.; w. at Gettysburg.
Vestal, H. H., e. May 28th, '61; Chatham co.
Vestal, Orren D., e. May 28th, '61; Chatham co.; w. Oct., '63, at Bristoe Station.
Ward, James, e. May 28th, '61; Chatham co.; k. at Gettysburg.
Ward, Stephen, e. May 28th, '61; Chatham co.; w. at Gettysburg.
Webster, Alexander, e. March, '62; Chatham co.; dt.
Welch, Adolphus A., e. May, '62; Chatham co.; w. March 2d, '62.
Welch, Henry B., e. Aug., '61; Chatham co.; w. at Gettysburg and Bristoe Station.
Welch, Levi B., e. May '61; Chatham co.; w. at Gettysburg.
Welch, Robert, e. August, '61; Chatham co.; w. at Gettysburg.
Welch, W. J., e. May, '61; Chatham co.; w. at Malvern Hill and Gettysburg.
Wilkey, D. F., e. May, '61; Chatham co.
Wilkey, Isaac H., e. March, '63; Chatham co.; w. at Gettysburg.
Williams, A. G., e. June 22d, '61; Chatham co.
Womble, Jesse C., e. May, '63; Chatham co.; d. May, '62, at Kinston.

COMPANY F.

OFFICERS.

Nathaniel P. Rankin, Captain, cm. July 15th, '61; Guilford co.; p. Major.
R. M. Tuttle, Captain, Caldwell co.; p. from 1st Lieut. and w.

Joseph R. Ballou, 1st Lieut., cm. July 15th, '61; Caldwell co.
R. M. Tuttle, 1st Lieut., cm. April 1st, '62; Caldwell co.
John B. Holloway, 2d Lieut., cm. July 15th, '61; Caldwell co.
Alfred T. Stewart, 2d Lieut., cm. July 15th, '61; Caldwell co.
C. M. Sudderth, 2d Lieut., cm. April 12th, '62; Caldwell co.
R. W. Hudspeth, 2d Lieut., cm. Sept. 1st, '63; Caldwell co.
A. B. Hayes, 2nd Lieut., cm. April 12th, '62; Caldwell co.

NON-COMMISSIONED OFFICERS.

R. M. Tuttle, 1st Sergeant, e. July 15th, '61; Caldwell co.; p. 1st Lieutenant April 1st, '62.

H. N. Coffey, 2d Sergeant, e. July 15th, '61; Caldwell co; dg. May 22d, '62.

J. T. C. Hood, 3d Sergeant, e. July 15th, '61; Caldwell co.; p. and w. at Gettysburg.

O. C. Collette, 4th Sergeant, e. July 16th, '61; Caldwell co.; tr. to 58th Regiment.

Caleb Estes, 1st Corporal, e. July 15th, '61; Caldwell co.; p. 1st Sergt. Sept. 1st, '62, and pr. at Gettysburg.

S. P. Philyans, 2d Corporal, e. July 15th, '61; Caldwell co.; w. July 1st, '63, at Gettysburg.

James A. White, 3d Corporal, e. July 15th, '61; Caldwell co.; w. July 1st, '62, and tr. to. Com. E, 58th Regiment.

George L. Powell, 4th Corporal, e. July 15th, '61; Caldwell co.; pr.

PRIVATES.

Arney, George, e. January 7th, '63; Caldwell co.; w. July 1st, '63, at Gettysburg.

Arney, Eli, e. August 20th, '63; Caldwell co.; w.

Annas, Hezekiah, e. January 5th, '61; Caldwell co.; w. July 1st, '63, at Gettysburg.

Anton, W. B., e. August 20th, '61; Caldwell co.; w. Oct. 14th, '63, at Bristoe Station.

Abshur, Thomas, e. Nov. 9th, '63; Caldwell co.

Abshur, John, e. Nov. 9th, '63; Caldwell co.

Brown, Leander, e. July 15th, '61; Caldwell co.; d. May 14th, '62, at Kinston.

Barnett, John, e. July 15th, '61; Caldwell co.; dg. April, '62.

Bradshaw, Nathan, e. July 15th, '61; Caldwell co.

Barker, Davis, e. July 15th, '61; Caldwell co.

Barber, W. M., e. July 15th, '61; Caldwell co.; p. Sergeant September, '61.

Barber, D. W., e. June 15th, '61; Caldwell co.

Barber, Thomas M., e. October 1st, '62; Caldwell co.; tr. from Com. E, 58th Regiment.

Beach, Zero, e. July 15th, '61; Caldwell co.; w. July 1st, '63, at Gettysburg; pr. in '64.

Braswell, L. M., e. August 9th, '61; Caldwell co.; d. August 28th, '62, at Petersburg.

Braswell, James, e. March 20th, '62; Caldwell co.

Braswell, T. N., e. October 15th, '62; Caldwell co.

Braswell, R. W., e. October 15th, '62; Caldwell co.

Braswell, R. M., e. October 15th, '62; Caldwell co.; k. July 3d, '63, at Gettysburg.

Braswell, M. L., e. October 15th, '62; Caldwell co.

Bowman, John, e. March 20th, '62; Caldwell co.; p. Corporal September, '62; w. at Gettysburg.

Bowman, J. B., e. October 18th, '61; Caldwell co.; d. May 2d, '62, at Kinston.

Bowman, J. W., e. March 20th, '62; Caldwell co.; d. May 20th, '62, at Kinston.

Bowman, Robertson, e. August, '63; Caldwell co.

Bowman, Hosea, e. March 20th, '62; Caldwell co.; dg. May 1st, '62.

Bowman, W. R., e. August 20th, '63; Caldwell co.; d. in '64.

Bradford, Hosea, e. June 22d, '62; Caldwell co.

Bradford, J. B., e. March 20th, '62; Caldwell co.; d. August 6th, '62, of w. received at Malvern Hill.

Bradford, E. J., e. August 9th, '61; Caldwell co.; d. Feb. 9th, '63, at Petersburg.

Bradford, William W., e. January 7th, '63; Caldwell co.; w. at Gettysburg.

Bean, John J., e. March 20th, '62; Caldwell co.; dt.

Bean, W. W., e. March 20th, '62; Caldwell co.; w. at Gettysburg and Bristoe Station.

Badger, S. P., e. October 18th, '61; Caldwell co.; w. at Gettysburg.

Branch, Jackson, e. March 20th, '62; Caldwell co.; dt. Dec., '62; pr. in '64.

Baldwin, Joseph, e. February 20th, '63; Caldwell co.; w. at Gettysburg.

Blaylock, L. M., e. March 20th, '62; Caldwell co.; dg. April 20th, '62.

*Blaylock, Mrs. L. M., e. March 20th, '62; Caldwell co.; dg. for being a woman.

Coffey, H. C., e. July 15th, '61; Caldwell co.; p. Sergeant; w. at Gettysburg.

Coffey, J. G., e. July 15th, '61; Caldwell co.; d. at David's Island August 26th, '63, of w. received at Gettysburg.

Coffey, J. P., e. July 15th, '61; Caldwell co.

Coffey, David, e. July 15th, '61; Caldwell co.; d. Dec. 23d, '61, at Morehead City.

Coffey, J. H., e. October 1st, '61; Caldwell co.; k. July 1st, '63, at Gettysburg.

Coffey, Cleveland, e. March 20th, '62; Caldwell co.; d. July 3d, '63, of w. received July 1st, '63.

Coffey, G. W., e. March 20th, '62; Caldwell co.

Coffey, T. M., e. March 20th, '62; Caldwell co.; d. August 12th, '63, of w. received at Gettysburg.

Coffey, W. S., e. March 20th, '62; Caldwell co.; d. August 20th, '63, of w. received at Gettysburg.

Coffey, Larkin, e. February 4th, '63; Caldwell co.

Coffey, J. A., e. March 20th, '62; Caldwell co.; w. July 1st, '63, at Gettysburg.

Coffey, N. C., e. March 20th, '62; Caldwell co.; tr. to Com. E, 58th Regiment.

Coffey, John, e. March 20th, '62; Caldwell co.; d.. April 15th, '62.

Coffey, Irwin, e. March 20th, '62; Caldwell co.; tr. to Com. H, 58th Regiment.

Coffey, Armistead, e. March 20th, '62; Caldwell co.; tr. to Com. H, 58th Regiment.

Coffey, James F., e. March 20th, '62; Caldwell co.; d. Aug. 15th, '62, at Petersburg.

Caswell, R. H., e. January 5th, '63; Burke co.; k. at Gettysburg July 1st, '63.

Carroll, James, e. July 15th, '61; Burke co.; dg. August 20th, '61.

Curtis, Joshua, e. July 15th, '61; Caswell co.

Curtis, William, e. December 1st, '62; Caswell co.; tr. from Company E, 58th Regiment; w. at Gettysburg.

Curtis, Thomas, e. October 10th, '61; Caswell co.; w. July 1st, '63, at Gettysburg.

Courtney, H. O., e. July 15th, '61; Caswell co.; w. July 1st, '63, at Gettysburg.

Courtney, A. H., e. July 17th, '61; Caswell co.; p. Corporal; w. at Gettysburg.

Clouts, Mortimer, e. July 15th, '61; Caswell co.; d. November 8th, '63, of w. received at Gettysburg.

Cozort, Thomas J., e. July 15th, '61; Caswell co.; dt. Pioneer, and k. at Gettysburg.

Corpening, D. M., e. July 15th, '61; Burke co.; dg. May 1st, '62.

Corpening, I. N., e. October 14th, '61; Caldwell co.; p. 1st Sergeant April, '62.

Clarke, Joseph C., e. March 20th, '62; Caldwell co.; w. July 1st, '63, at Gettysburg; pr. in '64.

Clarke, William, e. January 5th, '63; Burke co ; w. at Gettysburg in three places.

Culbreath, Nathaniel, e. March 20th, '62; Caldwell co.; w. at Gettysburg.

Crisp, S. W., e. March 20th, '62; Caldwell co.; w. at Gettysburg; pr. in '64.

Church, Redmond, e. March 20th, '62; Caldwell co.

Carraway, Giles, e. March 20th, '62; Caldwell co.

Crump, Thomas, e. March 20th, '62; Caldwell co.; d. near Martinsburg July 10th, '63, of w. received at Gettysburg.

---

*This lady had done a soldier's duty without a suspicion of her sex among her comrades, until her husband, L. M. Blaylock, was discharged, when she claimed the same privilege, and was sent home rejoicing.

Crump, H. C., e. March 20th, '62; Caldwell co.

Crump, J. M., e. March 20th, '62; Caldwell co; w. October 14th, '63, at Bristoe Station; p. Corporal; pr. in '64.

Crump, John, e. January 7th, '63; Caldwell co.; d. October 18th, '63, of w. received at Bristoe Station.

Campbell, John, e. July 18th, '61; Caldwell co.; p. Sergeant and dg. May 1st, '62.

Cannon, Wesley, e. October 1st, '61; Caldwell co.; tr. from Com. H, 35th Regiment.

Cook, H. H., e. October 14th, '61; Caldwell co.; d. May 18th, '62, at Kinston.

Calloway, Giles, e. March 20th, '62; Caldwell co.; pr. in '64.

Danner, David, e. March 20th, '62; Caldwell co.

Deal, James, e. January 20th, '63; Caldwell co.

Davis, James, e. August 20th, '63; Caldwell co.; d. in '64.

Davis, William H., e. July 15th, '61; Caldwell co.; d. September 1st, '63, at home.

Estis, William, e. July 15th, '61; Caldwell co.; w. July 1st, '62, at Malvern Hill.

Estes, D. W. P., e. July 15th, '61; Caldwell co.; dg. May 1st, '62.

Estes, L. C., e. July 15th, '61; Caldwell co.; dg. May 1st, '62.

Estes, J. J., e. March 20th, '62; Caldwell co.; tr. to Com. E, 58th Regiment.

Ephland, William, e. October 28th, '62; Randolph co.; d. Feb. 20th, '63, at Petersburg.

Ervin, Rufus, e. January 5th, '63; Burke co.; w. at Gettysburg.

Fletcher, James, e. July 15th, '61; Caldwell co.; dg. December 1st, '61.

Fleming, J. W., e. March 20th, '62; Caldwell co.; d. April 20th, '62, at Greenville.

Fleming, William, e. January 7th, '63; Caldwell co; k. at Gettysburg.

Furr, Abram, e. March 20th, '62; Caldwell co.; d. Sept. 1st, '62, at Raleigh.

Friddle, George, e. March 20th, '62; Caldwell co.; d. Sept. 5th, '62, at Petersburg.

Ford, John, e. January 15th, '63; Watauga co.

Gilbert, Vestal, e. July 15th, '61; Caldwell co.; dg. December 15th, '62.

Gilbert, James, e. March 20th, '62; Caldwell co.

Gilbert, Osman, e. February 1st, '63; Caldwell co.; tr. from Com. E, 58th Regiment, February 1st, '63.

Gragg, Jackson, e. March 20th, '62; Caldwell co.; k. at Gettysburg.

Holloway, G. W., e. July 15th, '61; Caldwell co.; p. Sergeant; w. at Malvern Hill and Gettysburg; pr. in '64.

Holloway, J. M., e. July 15th, '61; Caldwell co.

Holloway, J. P., e. March 20th, '62; Caldwell co.; d. July 3d, '62, at Richmond.

Hood, G. W., e. July 15th, '61; Caldwell co.; d. December 1st, '63, of w. received at Bristoe Station.

Hairston, G. H., e. July 15th, '61; Caldwell co.; w. at Bristoe Station.

Hudspeth, A. M., e. July 15th, '61; Caldwell co; k. Oct. 14th, '63, at Bristoe Station.

Hudspeth, G. W., e. July 17th, '61; Caldwell co.; w. July 1st, '63, at Gettysburg.

Hudspeth, R. N., e. July 17th, '61; Caldwell co.; p. Corporal Nov., '62; Sergeant Dec. 1st, '62; 2d Lieut. Sept. 1st, '63; d. in '64.

Hamell, Paul, e. July 17th, '61; Caldwell co.; w. July 1st, '63, at Gettysburg; pr. in '64.

Hayes, A. B., e. July 17th, '61; Caldwell co.; p. 2d Lieut. April 12th, '62.

Hayes, John L., e. July 17th, '61; Caldwell co.; dg. May 22d, '62.

Hayes, H. H., e. March 20th, '62; Caldwell co.; w. July 1st, '63, at Gettysburg.

Hayes, John W., e. March 20th, '62; Caldwell co.; d. July 27th, '63, at Petersburg.

Hart, Pinkey, e. July 15th, '61; Caldwell co.; d. April, '62, at home.

Harris, J. C., e. August 20th, '63; Caldwell co.

Hudson, Abram, e. October 20th, '62; Burke co.; k. July 1st, '63, at Gettysburg.

Huie, Benjamin, e. July 15th, '61; Caldwell co.; dg. December 1st, '61.

Hudson, Ambrose, e. October 20th, '62; Burke co.

Hicks, James, e. January 12th, '63; Caldwell co.

Icenhour, Philip, e. March 20th, '63; Caldwell co.; d. Dec. 12th, '62, at Petersburg.

Justice, John, e. July 15th, '61; Caldwell co.; tr. to Com. I March, '63.

Jones, Floyd T., e. March 20th, '62; Caldwell co.; p. 1st Sergeant June 1st, '62.

Jones, E. W., e. September 15th, '62; Wilkes co.; tr. from Com. F, 52d Regiment.

Kirby, A. P., e. July 15th, '61; Caldwell co.; w. October 14th, '63, at Bristoe Station.

Kirby, W. N., e. July 15th, '61; Caldwell co.; p. Corporal October, '63; w. at Gettysburg.

Kirby, W. P., e. March 20th, '62; Caldwell co.; w. October 14th, '63, at Bristoe Station.

Kirby, M. R., e. March 20th, '62; Caldwell co.; dg. July 2d, '62.

Kincaid, H. O., e. July 15th, '61; Burke co.; d. May 17th, '62, at Kinston.

Kincaid, John, e. March 20th, '62; Caldwell co.; w. July 1st, '63, at Gettysburg.

Keller, Thompson, e. March 20th, '62; Caldwell cc.

Lewis, J. C., e. July 15th, '61; Caldwell co.; k. at Gettysburg.

Lewis, W. A. G., e. March 20th, '62; Caldwell co.; pr. in '64.

Littlejohn, J. B., e. July 18th, '63; Caldwell co.; d. July 3d, '63, of w. received at Gettysburg.

Littlejohn, T. F., e. August 20th, '63; Caldwell co.

Littlejohn, P. W., e. February 14th, '64; Caldwell co.

L'Argent, Philip, e. March 20th, '62; Caldwell co.

Morgan, George, e. July 15th, '61; Caldwell co.; d. August 4th, '63, of w. received at Gettysburg.

McCarver, John, e. July 15th, '61; Caldwell co.

Moore, James D., e. July 15th, '61; Caldwell co.; w. at Gettysburg.

Mathis, Elkanah, e. February 4th, '63; Caldwell co.; w. at Gettysburg.

Miller, Henry, e. March 20th, '62; Caldwell co.; d. July 21st, '62, at Richmond.

Maske, Dudley, e. March 20th, '62; Caldwell co.; tr. to Com. H, 58th Regiment

Nelson, A. J., e. July 15th, '61; Caldwell co.; dg. Sept. 1st, '62.

Nelson, J. H., e. Nov. 6th, '61; Caldwell co.; p. Corporal and Sergeant; w. at Rolles' Mill and pr. in '64.

Pless, R. R., e. July 15th, '61; Burke co.; d. Sept. 25th, '62, at home.

Perkins, A. W., e. July 15th, '61; Caldwell co.; w. July 1st, '63, at Gettysburg.

Porch, George, e. July 15th, '61; Caldwell co.; w. July 1st, '63, at Gettysburg.

Porch, John, e. Dec. 14th, '62; Caldwell co.; w. July 1st, '63, at Gettysburg.

Powell, Pinkney, e. Feb. 20th, '63; Caldwell co.; w. July 1st, '63, at Gettysburg.

Page, Noah, e. Jan. 5th, '63; Burke co.; w. July 1st, '63, at Gettysburg.

Pergurson, R., e. October 28th, '62; Randolph co.

Philyan, Gideon, e. July 15th, '61; Caldwell co.

Pendley, James, e. March 20th, '62; Caldwell co.; dg. Sept. 15th, '62.

Pendley, Silas, e. March 20th, '62; Caldwell co.; dg. Dec. 1st, '62.

Phillips, Joseph, e. March 20th, '62; Caldwell co.; k. July 1st, '63, at Gettysburg.

Phillips, W. E., e. March 20th, '62; Caldwell co.; k. July 1st, '63, at Gettysburg.

Payne, Wm. R., e. Nov. 1st, '61; Caldwell co.; tr. from Com. I.

Page, E., e. October 20th, '61; Burke co.; d. Feb. 25th, '63, at Weldon.

Powell, D. N., e. July 15th, '61; Caldwell co.; dg. July 16th, '62.

Prestwood, E., e. October 8th, '61; Caldwell co.

Prestwood, C., e. August 25th, '63; Caldwell co.; k. October 14th, '63, at Bristoe Station.

Rainey, J. W., e. July 15th, '61; Caldwell co.; dg. May 28th, '63.

Rader, Jonas, e. July 15th, '61; Caldwell co.; p. Sergeant and d. April 10th, '63, at Goldsboro.

Rader, M. M., e. March 20th, '62; Caldwell co.; w. July 1st, '63, at Gettysburg.

Rich, W. B., e. August 9th, '61; Caldwell co.; w. at Gettysburg and Bristoe Station.

Rich, W. H., e. August 9th, '61; Caldwell co.; w. at Gettysburg.

Setser, W. H., e. July 15th, '61; Caldwell co.; d. July 4th, '63, of w. received at Gettysburg.

Setser, T. W., e. August 9th, '61; Caldwell co.; p. Sergeant; w. July 1st, '63, at Gettysburg.

Setser, T. D., e. March 20th, '62; Caldwell co.; d. May 20th, '62, at Kinston.

Setser, Joseph, e. September 8th, '62; Caldwell co.; d. July 17th, '63, of w. received at Gettysburg.

Setser, Adam, e. January 10th, '64; Caldwell co,

Swanson, S. N., e. October 8th, '64; Caldwell co.

Shell, J. W., e. July 5th, '62; Caldwell co.

Sumter, W. L., e. July 15th, '61; Caldwell co.; dg. January 1st, '62,

Sumter, John, e. March 20th, '62; Caldwell co.; pr. in '64.

Scott, Larkin, e. July 15th, '61; Caldwell co.

Shook, J. P., e January 5th, '63; Burke co.; k. at Gettysburg July 3rd, '63.

Stallings, Wiley, e. August 9th, '61; Caldwell co.; d. May 8th, '62, at Kinston.

Stallings, Hosea, e. August 9th, '61; Caldwell co.; d. of w. received at Gettysburg.

Stallings, William, e. March 9th, '61; Caldwell co.; w. at Gettysburg.

Sudderth, J. M., e. March 12th, '63; Caldwell co.; tr. from 58th Regiment, and w. July 1st, '63, at Gettysburg.

Sudderth, T. F., e. January 12th, '63; Caldwell co.; tr. from 58th Regiment; w. at Gettysburg and Bristoe Station.

Sudderth, C. M., e. March 20th, '62; Caldwell co.; p. 2nd Lieut. April 22nd, '62; d. January 25th, '65.

Suttlemire, A., e. January 5th, '62; Burke co.

Sherrill, Elisha, e. July 18th, '61; Caldwell co.

Sherrill, G. D., e. July 16th, '61; Caldwell co.; p. Corporal.

Stafford, Julius, e. July 18th, '61; Caldwell co; dg. August 20th, '61.

Taylor, A. I., e. July 18th, '61; Caldwell co.; k. July 1st, '63, at Gettysburg.

Taylor, J. S., e. March 20th, '62; Caldwell co.; d. May 12th, '62, at Kinston.

Taylor, George, e. July 15th, '61; Caldwell co.; dg. May 1st, '62.

Taylor, Benjamin, e. August 9th, '61; Caldwell co.; pr. in '64.

Taylor, Robert, e. July 15th, '61; Caldwell co.; pr. in '64.

Taylor, Zachary, e. January 10th, '64; Caldwell co.

Taylor, R. V., e. October 8th, '64; Caldwell co.

Towell, S. P., e. February 1st, '63; Caldwell co.

Tuttle, C. A., e. Feb. 1st, '63; Caldwell co.

Tuttle, C. A., e. December 17th, '61; Caldwell co.; w. at Gettysburg and Bristoe Station.

Tuttle, J. A., e. July 15th, '61; Caldwell co.; p. Sergeant; k. October 14th, '63, at Bristoe Station.

Teague, Logan, e. July 15th, '61; Caldwell co.; w. October 14th, '63, at Bristoe Station.

Townsell, William P., e. July 15th, '61; Caldwell co.

Townsell, M. L., e. July 15th, '61; Caldwell co.; k. July 1st, '63, at Gettysburg.

Thomas, S. A., e. December 2d, '61; Caldwell co.

Thomas, L. A., e. March 20th, '62; Caldwell co.; w. at Gettysburg.

Thompson, J. C., e. March 20th, '62; Caldwell co.; w. at Gettysburg.

Thompson, W. L., e. March 20th, '62; Caldwell co.; k. at Gettysburg.

Underdown, J. W., e. July 18th, '61; Caldwell co.; w. at Gettysburg; d. in '64.

Underdown, William, e. March 20th, '62; Caldwell co.; k. at Gettysburg.

Upchurch, Richard, e. July 15th, '61; Caldwell co.; w. at Gettysburg.

Upchurch, W. H., e. March 20th, '62; Caldwell co.; pr. in '64.
Woodruff, Noah, e. July 15th, '61; Caldwell co.; dg. December 1st, '61.
Wood, S. P., e. November 20th, '62; Caldwell co.
Winkler, Joseph, e. November 20th, '62; Caldwell co.; w. July 1st, '63, at Gettysburg.
White, Joseph, e. July 15th, '61; Caldwell co.; dg. August 25th, '61.
Zimmerman, Israel, e. January 5th, '63; Burke co.; w. at Gettysburg.

COMPANY G.

OFFICERS.

W. S. McLean, Captain, cm. June 10th, '61; Chatham co.
John R. Lane, Captain, cm. April 12th, '61; Chatham co.; p. Major, Lieut. Col., and Colonel; w.
A. R. Johnson, Captain, Chatham co.; p. from 1st Lieut.

John E. Matthews, 1st Lieut., cm. April 12th, '62; Chatham co.
A. R. Johnson, 1st Lieut , Chatham co.; p. from 2nd Lieut.
W. G. Lane, 1st Lieut., Chatham co.
George C. Underwood, 2d Lieut., cm. April 12th, '62; Chatham co.
Henry C. Allwright, 2d Lieut., cm. April 12th, '61; Chatham co.
J. A. Lowe, 2d Lieut., cm. November 22d, '61; Chatham co.
A. R. Johnson, 2d Lieut., cm. July 23d, '62; Chatham co.
W. G. Lane, 2d Lieut., cm. Oct. 11th, '63; Chatham co.
Samuel E. Teague, 2d Lieut.; cm. in '64; Chatham co.; p. from Sergeant.

NON-COMMISSIONED OFFICERS.

J. A. Lowe, 1st Sergeant, e. June 10th, '61; Chatham co.; p. 2d Lieut. November 22d, '61.
A. R. Johnson, 2d Sergeant, e. July 10th, '61; Chatham co.; p. 2d Lieut. July 23d, '62.
T. A. Matthews, 3d Sergeant, e. July 10th, '61; Chatham co.; w. July 1st, '62, at Malvern Hill.
S. E. Teague, 4th Sergeant, e. July 10th, '61; Chatham co.; p. 2d Lieut. in '64; pr.
D. C. Murchison, 5th Sergeant, e. July 10th, '61; Chatham co.; w. and c. July 1st, '63, at Gettysburg.
John R. Lane, 1st Corporal, e. July 10th, '61; Chatham co.; p. and w.
W. P. Kirkerson, 2d Corporal, e. July 10th, '61; Chatham co.; d. July 2d, '63, of w. received July 1st, '63, at Gettysburg.
John A. J. Moran, 3d Corporal, e. July 10th, '61; Chatham co ; d. September 22d, '63, of w. received July 1st, '63, at Gettysburg.
D. H. Edwards, 4th Corporal, e. June 10th, '61; Chatham co.; p. Sergeant; k. July 1st, '63, at Gettysburg.
J. J. Record, Musician, e. June 10th, '61; Chatham co.; p. 3d Corporal September 1st, '63.
Jesse L. Smith, Musician, e. June 10th, '61; Chatham co.; p. 2d Sergeant September 1st, '63.

PRIVATES.

Allred, Osborne, e. June 10th, '61; Chatham co.
Allred, W. B., e. June 10th, '61; Chatham co.
Allred, Isaac, e. June 10th, '61; Chatham co.; w. July 1st, '63, at Gettysburg.
Adcock, John C., e. June 10th, '61; Chatham co.; w. July 3d, '63, at Gettysburg.
Allen, Nathan, e. June 10th, '61; Chatham co.
Allridge, William, e. March 5th, '63; Chatham co.; k. July 1st, '63, at Gettysburg.
Allred, Yancey, e. April 20th, '63; Chatham co.
Bowdoin, C. M., e. June 10th, '61; Chatham co,; k. July 1st, '63, at Gettysburg.
Bray, William L., e. June 10th, '61; Chatham co.
Bray, W. F., e. June 10th, '62; Chatham co.; tr. to Com. E, 44th Regiment.
Bridges, George R., e. June 10th, '61; Chatham co.
Branson, William M., e. June 10th, '61; Chatham co.
Branson, Eli, e. June 10th, '61; Chatham co.
Bowdoin, Jesse W., e. March 6th, '62; Chatham co.
Bray, Jasper N., e. March 6th, '62; Chatham co.
Buckner, Richard, e. March 6th, '62; Chatham co.; k. July 1st, '62, at Malvern Hill.
Bridges, J. L., e. March 6th, '62; Chatham co.; w. July 1st, '63, at Gettysburg.
Burke, A. J., e. March 6th, '62; Chatham co.
Burke, R. M , e. October 24th, '64; Wake co.
Burgess, J. C., e. October 24th, '64; Wake co.
Brown, Marshall, e. March 6th, '62; Chatham co.; k. July 1st, '63, at Gettysburg.
Brown, John D., e. June 10th, '61; Chatham co.
Brown, Lewis S., e. October 24th, '64; Wake co.
Brown, James C., e. May 1st, '62; Chatham co.
Brown, Henry C., e. Nov. 17th, '64; Wake co.
Black, Ephraim, e September 10th, '64; Wake co.
Baines, Rufus, e. September 10th, '62; Wake co.
Culbertson, J. W., e. June 1st, '64; Wake co.
Carter, W. G., e. June 10th, '61; Chatham co.; p. Corporal Dec., '62; w. July 1st, '62, at Malvern Hill, and July 1st, '63, at Gettysburg; p. Sergeant in '65.
Carter, S. S., e. June 10th, '61; Chatham co.; tr. to Partisan Rangers.
Carter, J. Q. A., e. June 10th, '61; Chatham co.; dg. May, '62.
Carter, Sam'l H., e. March 9th, '63; Chatham co.; w. July 1st, '63, at Gettysburg.
Dorset, Mathias, e. June 10th, '61; Chatham co.
Dorset, Siler, e. June 10th, '61; Chatham co.
Deakin, John A., e. June 10th, '61; Chatham co.; dg.
Dowdy, John, e. October 24th, '61; Wake co.
Edwards, Wiley, e. June 10th, '61; Chatham co.
Edwards, Murphy, e. June 10th, '61; Chatham co.
Edwards, J. D., e. March 6th, '62; Chatham co.; w. July 1st, '63, at Gettysburg.
Edwards, S. B., e. July 19th, '62; Wake co.
Fogleman, Young R., e. June 10th, '61; Chatham co.; w. July 1st, '63, at Gettysburg.
Fogleman, Riley Y., e. June 10th, '61; Chatham co.
Fogleman, John A., e. March 6th, '62; Chatham co.; w. July, '63, at Gettysburg.
Fox, Fields D., e. Oct. 24th, '61; Wake co.
Fox, Alfred M., e. Sept. 21st, '62; Wake co.
Fox, Andrews G., e. June 10th, '61; Chatham co.; d. March 24th, '62.
Fox, James L., e. June 10th, '61; Chatham co.; d. Dec., '62.
Foushee, William, e. Sept. 20th, '62; Chatham co.
Garrett, Wm. A., e. June 10th, '61; Chatham co.; k. July 1st, '63, at Gettysburg.
Girton, Milton, e. August 20th, '61; Chatham co.

Gilbert, James A., e. Nov. 25th, '61; Chatham co.
Garrett, Henry B., e. Nov. 1st, '62; Chatham co.; k. July 1st, '63, at Gettysburg.
Gardner, Thomas, e. Sept. 10th, '62; Chatham co.; w. July 1st, '63, at Gettysburg.
Gainey, Wm. B., e. Oct. 17th, '63; Wake co.
Holstead, J. M., e. June 10th, '61; Chatham co.; w. July 1st, '63, at Gettysburg.
Hicks, Quinby, e. June 10th, '61; Chatham co.
Hickman, Daniel, e. June 10th, '61; Chatham co.; d. March, '62.
Hickman, Eli, e. Jan. 25th, '64; Wake co.
Holmes, John D., e. March 5th, '61; Chatham co.
Henshaw, Thomas, e. June 10th, '61; Chatham co.
Hicks, John H., e. Sept. 10th, '61; Chatham co.
Hobson, John, e. July 1st, '63; Chatham co. d. Sept. 3d, '63.
Jordan, W. A., e. June 10th, '61; Chatham co.; w. July 1st, '63, at Gettysburg.
Jordan, R. F., e. March 6th, '62; Chatham co.; w. July 1st, '63, at Gettysburg.
Jordan, Isaac, e. March 5th, '63; Chatham co.; w. July 1st, '63, at Gettysburg.
Johnson, W. P., e. March 6th, '62; Chatham co.; w. July 1st, '63, at Gettysburg.
Johnson, Thomas C., e. March 6th, '62; Chatham co.; w. July 1st, '63, at Gettysburg.
Johnson, Jesse, e. March 6th, '62; Chatham co.
Johnson, Hiram, e. June 10th, '61; Chatham co.; p. 4th Sergeant; w. July 1st, '62, at Malvern Hill.
Johnson, H. W., e. June 10th, '61; Chatham co.; w. July 1st, '63, at Gettysburg.
Johnson, W. F., e. August 12th, '62; Chatham co.; w. July 1st, '63, at Gettysburg.
Johnson, L., e. September 10th, '63; Chatham co.
Johnson, Henry C., e. Sept. 11th, '62; Chatham co.
Jones, John S., e. September 10th, '63; Chatham co.; w. July 1st, '63, at Gettysburg.
Jones, James, e. September 10th, '63; Chatham co.; w. July 1st, '63, at Gettysburg.
Kirkman, W. P., e. June 10th, '61; Chatham co.
Kirkman, H. C., e. June 10th, '61; Chatham co.; w. July 3d, '63, at Gettysburg; d. September 15th, '63.
Kirkman, J. E. B., e. September 10th, '62; Chatham co.
Lane, A. J., e. June 10th, '61; Chatham co.; p. Q. M. of Regiment Sept. 1st, '63.
Lane, W. G., e. September 1st, '61; Chatham co.; p. 1st Sergeant and 2d Lieut. October 10th, '63.
Lane, H. C., e. October 25th, '61; Chatham co.; dg. October, '62.
Lineberry, W. A., e. March 6th, '62; Chatham co.; w. July 1 t, '62, at Malvern Hill, and July 1st, '63, at Gettysburg.
Lineberry, A. N., e. March 6th, '62; Chatham co.
Mann, Fletcher A., e. October 24th, '61; Wake co.
McPherson, J. W., e. July 4th, '61; Chatham co.; dg. August, '62.
McPherson, I. R., e. June 10th, '61; Chatham co.
McPherson, David B., e. June 10th, '61; Chatham co.
McPherson, Isaac P., e. June 10th, '61; Wake co.
Moran, Samuel M., e. June 10th, '61; Chatham co.; w. July 1st, '63, at Gettysburg.
Moran, William, e. June 10th, '61; Chatham co.; dg. May, '62.
Marley, John R., e. June 10th, '61; Chatham co.; k. July 1st, '63, at Gettysburg.
Murchison, W. G., e. June 10th, '61; Chatham co.; w. July 1st, '63, at Gettysburg; p. Sergeant in '65.
Murchison, Duncan C., e. June 10th, '61; Wake co.
Matthews, A. M., e. Jan. 14th, '64; Wake co.
Marshall, Eli, e. Oct. 24th, '61; Wake co.
Moore, Larkin, e. March 6th, '62; Chatham co.; w. July 1st, '63, at Gettysburg.
McCoy, Alston G., e. March 6th, '62; Chatham co.; k. Oct. 14th, '63, at Bristoe Station.

Marley, John G. H., e. March 6th, '62; Chatham co.
Moran, George W., e. August 12th, '62; Chatham co.
McMasters, H. M., e. October 24th, '61; Wake co.
Nelson, J. W., e. June 10th, '61; Chatham co.; w. July 1st, '63, at Gettysburg.
Nelson, Nathan, e. June 10th, '61; Chatham co.
Nelson, W. P., e. June 10th, '61; Chatham co.
Norwood, Nathaniel, e. March 6th, '62; Chatham co.; k. July 1st, '63, at Gettysburg.
Norwood, Gilliam, e. June 10th, '61; Chatham co.; w. July 1st, '61, at Gettysburg.
Norwood, Washington, e. Sept. 10th, '61; Chatham co.; w. July 1st, '63, at Gettysburg.
Norwood, G. W., e. Sept. 10th, '62; Chatham co.
Neese, Riley, e. March 6th, '62; Chatham co.
Overman, Hezekiah, Oct. 14th, '64; Wake co.
Overman, Wm., e. June 10th, '61; Chatham co.
Peevitt, Thomas, e. June 10th, '61; Chatham co.; p. Corporal and w. at Gettysburg.
Parish, Elias, e. June 10th, '61; Chatham co.; w. July 1st, '63, at Gettysburg.
Pervis, James L., e. June 10th, '61; Chatham co.
Perry, Thomas L., e. October 24th, '64; Wake co.
Patterson, W. H., e. June 10th, '61; Chatham co.
Pike, John, e. June 10th, '61; Chatham co.; w. July 1st, '63, at Gettysburg.
Parrish, John, e. March 6th, '61; Chatham co.
Patterson, J. H., e. Sept. 16th, '62; Chatham co.; w. July 1st, '63, at Gettysburg.
Poe, B. F., e. March 6th, '62; Chatham co.; tr. to Com. I, 22d Regiment.
Richardson, Noah, e. June 10th, '61; Chatham co.
Richardson, Samuel, e. January 10th, '64; Wake co.
Raines, W. R., e. June 10th, '61; Chatham co.
Record, David, e. June 10th, '61; Chatham co.; w. July 1st, '63, at Gettysburg.
Record, Thomas D., e. June 10th, '63; Chatham co.; p. Sergeant and w. July 1st, '63, at Gettysburg.
Rosson, C. F., e. June 10th, '61; Chatham co.
Rosson, Merrill, e. June 10th, '61; Chatham co.
Rogers, Jasper, e. June 10th, '61; Chatham co.
Reeves, W. H. C., e. June 10th, '61; Chatham co.; d. Sept. 22d, '63, in N. C., of w. received July 1st, '63, at Gettysburg.
Rousley, William, e. June 10th, '61; Chatham co.; dg.
Rightman, Alfred, e. June 10th, '61; Chatham co.; dg.
Roberts, J. H., e. May 6th, '62; Chatham co.
Rightsell, Samuel, e. September 10th, '62; Chatham co.; w. July 1st, '63, at Gettysburg.
Reitzel, John, e. Oct. 24th, '64; Wake co.
Rosson, Jas. F., e. January 1st, '63; Chatham co.; w. July 1st, '63, at Gettysburg.
Stuart, Levi, e. January 10th, '64; Wake co.
Silton, J. W., e. August 15th, '62; Chatham co.; dg. December, '62.
Smith, Solomon D., e. June 10th, '61; Chatham co.
Smith, J. B., e. June 10th, '61; Chatham co.; dg.
Sheridan, James M., e. March 6th, '62; Chatham co.; k. October 14th, '63, at Bristoe Station.
Smotherly, Wm., e. March 6th, '62; Chatham co.; w. July 1st, '63, at Gettysburg.
Siler, W. H., e. March 6th, '62; Chatham co.; w. July 1st, '63, at Gettysburg.
Smith, W. A., e. September 10th, '62; Chatham co.
Siler, J. Q. A., e. September 10th, '62; Chatham co.; w. July 1st, '63, at Gettysburg.
Siler, W. M., e. August 10th, '62; Chatham co.
Siler, A. R., e. August 15th, '63; Chatham co.

Siler, Edwin, e. June 10th, '61; Chatham co.; dg.
Siler, James L., e. June 10th, '61; Chatham co.
Siler, S. W. C., e. June 10th, '61; Chatham co.; d. July 12th, '62, at Richmond.
Siler, S. F., e. June 10th, '61; Chatham co.; d. Oct., '61, at Carolina City.
Siler, James M., e. September 10th, '62; Chatham co.; d. Nov., '62, at Petersburg.
Siler, Newlan J., e. Dec. 7th, '64; Wake co.
Teague, Willis, e, June 10th, '61; Chatham co.
Terry, W. H., e. August 12th, '62; Chatham co.; w. July 1st, '63, at Gettysburg.
Tighlman, Jarratt, e. September 10th, '62; Chatham co.; w. July 1st, '63, at Gettysburg.
Thompson, Isham, e. September 10th, '62; Chatham co.
Vestal, Henry T., e. March 6th, '62; Chatham co.
Vestal, Riley, e. June 10th, '61; dg. December, '61.
Vestal, W. J., e. July 10th, '62; Chatham co.; dg.
Vestal, Murphy, e. Feb. 1st, '64; Wake co.
Vinson, George W., e. June 10th, '61; w. July 1st, '63, at Gettysburg.
Vinson, John, e. June 10th, '61.
Wicker, Lennis, e. June 10th, '61; Chatham co.; k. July 1st, '63, at Gettysburg.
Way, A. A., e. June 10th, '61; Chatham co.
Wells, Anderson, e. June 10th, '61; Chatham co.
Way, Anderson, e. March 6th, '62; Chatham co.; k. July 1st, '63, at Gettysburg.
Wards, John, e. May 30th, '63; Chatham co.
Williams, E. R., e. Aug. 11th, '62; Guilford co.; tr.

COMPANY H.

OFFICERS.

William P. Martin, Captain, cm. June 30th, '61; Moore co.

Clement Dowd, 1st Lieut., cm. June 3d, '61; Moore co.; p. Major and A. Q. M.
James D. McIver, 2d Lieut., cm. June 3d, '61; Moore co.
Robert W. Goldston, 2d Lieut., cm. June 3d, '61; Moore co.
John H. McGilvary, 2d Lieut., cm. June 24th, '62; Moore co.; r. Jan. 29th, '63.
Murdoch McLeod, 2d Lieut., cm. June 24th, '62; Moore co.

NON-COMMISSIONED OFFICERS.

M. D. McNeil, 1st Sergeant, e. June 3d, '61; Moore co.; dg. June 5th, '63.
Richard Sheet, 2d Sergeant, e. June 3d, '61; Moore co.; dg.
L. P. Tyson, 3d Sergeant, e. June 3d, '61; Moore co.; d. April, '62.
James A. McLeod, 4th Sergeant, e. June 3d, '61; Moore co.; k. July 1st, '63, at Gettysburg.
John McNeil, 1st Corporal, e. June 3d, '61; Moore co.; d. Feb. 28th, '62, at New Bern.
John H. McGilvary, 2d Corporal, e. June 3d, '61; Moore co.; p. 2d Lieut. June 24th, '62, and r. Jan., '63.
Francis McIntosh, 3d Corporal, e. June 3d, '61; Moore co.; tr. to 30th Regiment Sept., '61.
Malcolm Brewer, 4th Corporal, e. June 3d, '61; Moore co.; w. July 1st, '63, at Gettysburg.

PRIVATES.

Blue, Calvin, e. June 3d, '61; Moore co.; dg. in '62.
Blue, William, e. May 12th, '62; Moore co.; w. July 1st, '63, at Gettysburg.
Beck, James, e. March 10th, '62; Moore co.
Brewer, S. W., e. June 3d, '61; Moore co.
Brown, E. S., e. March 15th, '62; Moore co.; tr. to Ord. Department Nov., '62.
Brewer, W. D , e. March 15th, '62; Moore co.; k. July 1st, '63, at Gettysburg.
Brown, J. S., e. March 15th, '62; Moore co.; m. at Gettysburg.
Brown, Archibald, e. June 3d, '61; Moore co.; dt. Musician.
Brown, Page, e. November 8th, '62; Moore co.; w. October 14th, '63, at Bristoe Station.
Brady, Bradley, e. June 5th, '61; Moore co.; w. July 1st, '63, at Gettysburg.
Bailey, Daniel, e. November 8th, '62; Moore co.; w. July 1st, '63, at Gettysburg.
Buchan, J. A. J., e. June 5th, '61; Moore co.; p. Sergeant in '64.
Brown, N. R., e. March 15th, '62; Moore co.; k. March, '62, at New Bern.
Cox, Hugh B., e. June 5th, '61; Moore co.; dg. November, '62.
Cox, Green B., e. June 5th, '61; Moore co.; p. Sergeant in April; d. May 20th, '63.
Cox, Thomas, e. June 5th, '61; Moore co.; d. February 10th, '62, at Carolina City.
Currie, N. A., e. June 5th, '61; Moore co.; k. July 5th, '63, at Gettysburg.
Currie, L. A., e. June 5th, '61; Moore co.; p. Sergeant June 28th, '62; w. July 3d, '63, at Gettysburg.
Currie, L. W., e. June 5th, '61; Moore co.
Campbell, P. M., e. June 5th, '61; Moore co.; dg. November, '62.
Campbell, W. R., e. June 5th, '61; Moore co.
Caddell, A. S., e. June 5th, '61; Moore co.; w. October 14th, '63, at Bristoe Station.
Caddell, James N., e. June 5th, '61; Moore co.
Caddell, C. D., e. June 15th, '63; Moore co.
Cagle, E. S., e. June 5th, '61; Moore co.; dg. August 10th, '62.
Cagle, Matthew, e. March 15th, '62; Moore co.
Clark, Arch'd A., e. January 15th, '63; Moore co.; k. July 1st, '63, at Gettysburg.
Caddell, John L., e. June 3d, '61; Moore co.; dg. December, '62.
Cox, Robert, e. November, '62; Moore co.
Comer, Peter, e. October 1st, '62; Moore co.; d. December 20th, '62, at Petersburg.
Cook, W. R., e. June 3d, '61; Moore co.; d. June 21st, '63, at Petersburg.
Dowd, W. J. D., e. June 5th, '61; Moore co.; w. July 1st, '63, at Gettysburg.
Denson, C. E., e. October 21st, '61; Moore co.; p. Captain of Engineers.
Denson, W. T., e. June 5th, '61; Moore co.
Dunlop, A. M., e. June 5th, '61; Moore co.; w. July 1st, '63, at Gettysburg.
Dunlop, D. M., e. June 6th, '61; Moore co.
Davis, Dunston, e. June 6th, '61; Moore co.; dg.
Davis, W. H. H., e. June 6th, '61; Moore co.; p. Sergeant April, '62; d. July, '62, at Petersburg.
Deaton, Noah, e. June 6th, '61; Moore co.
Everett, Troy, e. June 3d, '61; Moore co.; d. April 25th, '62, at Kinston.
Fields, Richard, e. Nov. 27th, '64; Moore co.
Ferguson, D. C., e. June 3d, '61; Moore co.; w. July 1st, '62, at Malvern Hill; p. Corporal.
Fry, A. B., e. June 3d, '61; Moore co.; w. July 1st, '63, at Gettysburg.
Fry, N. L., e. June 3d, '61; Moore co.
Fry, George T., e. June 3d, '61; Moore co.; w. July 1st, '63, at Gettysburg.
Fry, J. C., e. March 15th, '62; Moore co.; d. October 15th, '63, of w. received October 14th, '63, at Bristoe Station.
Graham, D. A., e. June 3d, '61; Moore co.; w. July 1st, '63, at Gettysburg.
Graham, N. C., e. June 3d, '61; Moore co.

Graham, W. H., e. June 3d, '61; Moore co.; w. July 1st, '63, at Gettysburg.
Gilliam, J. D., e. June 5th, '61; Moore co.; p. Sergeant.
Harrison, E. C., e. June 3d, '61; Moore co.; w.
Hogan, Zacheus, e. June 3d, '61; Moore co.; w. July 1st, '63, at Gettysburg.
Hogan, T. J., e. October 1st, '63; Orange co.; w. July 1st, '63, at Gettysburg.
Hollingsworth, B. G., e. June 5th, '61; Moore co.; w. and pr. October 14th, '63, at Bristoe Station.
Hunsucker, H. W., e. Oct. 11th, '62; Moore co.
Hunsucker, Nelson, e. Oct. 11th, '62; Moore co.
Jackson, John A., e. June 3d, '61; Moore co.
Johnson, John A., e. Oct. 11th, '62; Moore co.; w. July 1st, '62, at Malvern Hill.
Johnson, Thos., e. Oct. 11th, '62; Moore co.; k. July 1st, '63, at Gettysburg.
Johnson, S. Y., e. June 3d, '61; Moore co.; w. July 1st, '63, at Gettysburg.
Jones, C. E., e. June 3d, '61; Moore co.; w. March 14th, '62, at New Bern.
Jordan, J. B., e. Aug. 10th, '62; Moore co.
Keith, John R., e. June 3d, '61; Moore co.; w. July 1st, '63, at Gettysburg; p. Corporal in '64.
Kendreth, A. J., e. June 3d, '61; Moore co.; d. Nov. 20th, '62, at Petersburg.
Kennedy, Enoch, e. March, '62; Moore co.; d. Aug. 6th, '62, at Petersburg.
Kelly, John B., e. June 3d, '61; Moore co.; w. July 1st, '62, at Malvern Hill; d. Aug. 11th, '62.
Kelly, Duncan, e. June 3d, '61; Moore co.; d. June 26th, '62, at Petersburg.
Keith, Hugh, e. June 3d, '61; Moore co.; d. April 26th, '62, at Wilson.
Lawhorn, J. J., e. June 3d, '61; Moore co.
Lewis, Benj., e Nov., '62; Moore co.
Martin, John B., e. June 3d, '61; Moore co.
Malone, Aaron, e. June 3d, '61; Moore co.; w. July 1st, '62, at Malvern Hill.
Malone, Wm., e. June 3d, '61; Moore co.; w. Oct. 14th, '63, at Bristoe Station.
McLeod, D. M., e. June 3d, '61; Moore co ; w. July 3d, '63, at Gettysburg.
McLeod, M., e. July 3d, '61; Moore co.; p. 2d Lieut. in '63.
Maness, Thos. S., e. June 3d, '61; Moore co.; w.
Maness, J. S., e. June 3d, '61; Moore co.; w. July 1st, '63, at Gettysburg.
Maness, J. W., e. March, '62; Moore co.
Maness, Isaac, e. March, '62; Moore co.
Maness, John L., e. March, '62; Moore co.
Maness, John, e. March, '62; Moore co.
McCallum, Archibald D., e. January, '63; Moore co.; w. July 1st, '63, at Gettysburg; d. in '64.
McCaskill, Daniel, e. June 3d, '61; Moore co.; w. July 1st, '63, at Gettysburg.
McDonald, Daniel, e. June 3d, '61; Moore co.; p. Corporal.
McDonald, W. H., e. June 3d, '61; Moore co.; w. Oct. 14th, '63, at Bristoe Station.
McKimmon, John, e. June 3d, '61; Moore co.; k. July 1st, '63, at Gettysburg.
McKimmon, William, e. June 3d, '61; Moore co.; w. July 1st, '63, at Gettysburg.
McKimmon, Martin, e. June 3d, '61; Moore co.; d. August 9th, '62, at Petersburg.
McKinnon, C. B., e. May, '62; Moore co.; k. July 1st, '63, at Gettysburg.
McNeill, W. J., e. June 3d, '61; Moore co.; k. October 14th, '63, at Bristoe Station.
McNeill, Archibald, Sr., e. July 10th, '62; Moore co.; p. Corporal April, '62.
McNeill, Archibald, Jr., e. November, '62; Moore co.
McDonald, Neill, e. August 3d, '61; Moore co.; p. Corporal; k. October 14th, '63, at Bristoe Station.
McRae, Alexander, e. November, '62; Moore co.
Moore, William E., e. June 3d, '61; Moore co.
Moore, Bryant H., e. June 3d, '61; Moore co.
Muse, Ashley F., e. March, '63; Moore co.; k. July 1st, '63, at Gettysburg.
McAuley, A. M., e. June 3d, '61; Moore co.; p. Sergeant and k. at Gettysburg.

McIntosh, Neill, e. June, '61; Moore co.; p. Sergeant April, '62.
Moffit, W. L., e. March, '62; Moore co.; d. August 10th, '62, at Petersburg.
McIntosh, S. J., e. June, '61; Moore co.; k. July 1st, '63, at Gettysburg.
Medlin, J. A., e. June 5th, '61; Moore co.; w. July 1st, '63, at Gettysburg.
Malone, Daniel, e. June 5th, '61; Moore co.; k. July 1st, '63, at Gettysburg.
Nunney, William R., e. June 3d, '61; Moore co.; w. July 1st, '63, at Gettysburg.
Nicholson, John, e. June 3d, '61; Moore co.; dg. November, '62.
Parsons, John, e. June 3d, '61; Moore co.; w. July 1st, '63, at Gettysburg.
Persons, William M., e. January, '63; Moore co.; d. August 2d, '63, of w. received at Gettysburg.
Patterson, William H., e. March, '62; Moore co.
Perry, Benj., e. November, '62; Moore co.
Ray, N. A., e. June 3d, '61; Moore co.
Roberts, C. C., e. June 3d, '61; Moore co.; w. July 1st, '63, at Gettysburg.
Ray, Hugh M., e. June 3d, '61; Moore co.; k. March 14th, '62, at New Bern.
Schermerhorn, I., e. June 3d, '61; Moore co.; dg. and d. Nov., '62.
Short, S. P., e. June 3d, '61; Moore co.; p. Corporal and k. at Gettysburg.
Seawell, Jesse P., e. June 3d, '61; Moore co.; w. at Gettysburg.
Seawell, Eli, e. March, '62; Moore co.
Seawell, Joseph, e. November, '62; Moore co.; w. July 1st, '63, at Gettysburg.
Street, Mindo, e. June, '61; Moore co.; dg. November, '62.
Stults, Lindsay, e. April, '62; Moore co.; w. at Gettysburg.
Stults, Andrew, e. November, '62; Moore co.; w. at Gettysburg.
Stults, B. W., e. November, '62; Moore co.
Shaw, Daniel M., e. June, '61; Moore co.; p. A. A. Surgeon; d. November 29th, '62, at Morehead City.
Shaw, Thomas B., e. June, '61; Moore co.; p. Ordnance Sergeant.
Shaw, C. W., e. March, '62; Moore co.; w. at Gettysburg and Bristoe Station.
Shields, N. P., e. November, '62; Moore co.
Stephens, J. W., e. March, '62; Moore co.
Sheffield, Isham, e. March, '62; Moore co.
Smith, Neill T., e. June, '61; Moore co.; k. October 14th, '63, at Bristoe Station.
Smith, N. P., e. June, '61; Moore co.; dg. in '62.
Smith, O. A., e. June, '61; Chatham co.; tr. to Company E August, '62.
Teague, William, e. November, '62; Moore co.; d. March 30th, '62, at Wilson.
Tyson, D. P., e. March, '62; Moore co.
Teague, Isaac, e. November, '62; Moore co.; d. December 15th, '62, at Petersburg.
Teague, Eli, e. March, '62; Moore co.
Tyson, H. C., e. March, '62; Moore co.; w. July 1st, '63, at Gettysburg.
Tyson, James, e. November, '63; Moore co.; w. October 14th, '63, at Bristoe Station.
Taylor, N. B., e. June, '61; Moore co.; dg. November, '62.
Tyson, Jordan, e. November, '62; Randolph co.; w. July 3d, '63, at Gettysburg.
Tyson, L. B., e. June, '61; Moore co.; d. March 17th, '62, of w. received at New Bern.
Thompson, Neill, e. June, '61; Moore co.; dg. December, '62.
Vuncannon, W. A., e. June, '61; Moore co.; w. at Gettysburg.
Wallace, William L., e. June, '61; Moore co.
Wallace, Alexander, e. June, '61; Moore co.
Wadsworth, Alexander, e. June, '61; Moore co.; dg. Nov., '62.
Wadsworth, P. S., e. June, '61; Moore co.
Whitlock, William, e. June, '61; Moore co.; dg. Nov., '62.
Warren, J. H., e. March, '62; Moore co.; w. July 1st, '63, at Gettysburg, and p. Sergeant in '64.
Warren, A. L., e. March, '62; Moore co.; k. October 14th, '63, at Bristoe Station.

Warwick, John T., e. June, '61; Moore co.
Williams, John, e. Nov., '62; Moore co.
Williams, Henry, e. Nov., '62; Moore co.; w. October 14th, '63, at Bristoe Station.
Williams, E., e. Nov., '62; Moore co.
Williams, A. P., e. March, '62; Moore co.; w. July 1st, '63, at Gettysburg.
Williams, A. J., e. March, '62; Moore co.; w. at Gettysburg and Bristoe Station.
Williams, A. M. D., e. March, '62; Moore co.; w. at Gettysburg and Bristoe Station and pr. in '64.
Williamson, Kelly, e. March, '62; Moore co.; w. July 1st, '63, at Gettysburg.
Williamson, N. R., e. March, '62; Moore co.
Wilcox, R. P., e. June 5th, '61; Moore co.; p. Sergeant.
Yow, Henry, e. Nov., '62; Moore co.; w. July 1st, '63, at Gettysburg.

COMPANY I.

OFFICERS.

Wilson A. White, Captain, cm. July 26th, '61; Caldwell co.
N. G. Bradford, Captain, cm. ——; Caldwell co.

John Catson, 1st Lieut., cm. July 26th, '61; Caldwell co.
S. P. Dula, 1st Lieut., cm. April 14th, '62; Caldwell co.
N. G. Bradford, 1st Lieut., cm. in '63; Caldwell co.
M. B. Blair, 1st Lieut., cm. in '64; Caldwell co.
John Jones, 2d Lieut., cm. July 26th, '61; Caldwell co.
M. B. Blair, 2d Lieut., cm. July 26th, '61; Caldwell co.
N. G. Bradford, 2d Lieut., cm. April 14th, '62; Caldwell co.
J. C. Grier, 2d Lieut., cm. September, '62; Caldwell co.
J. A. Bush, 2d Lieut., cm. ——; Caldwell co.

NON-COMMISSIONED OFFICERS.

N. G. Bradford, 1st Sergeant, e. July 26th, '61; Caldwell co.; p. 2d Lieutenant April 14th, '63.
J. B. Henderson, 2d Sergeant, e. July 26th, '61; Caldwell co.; pr. at Gettysburg.
J. W. Shell, 3d Sergeant, e. July 26th, '61; Caldwell co.; d. April, '62.
James Oxford, 4th Sergeant, e. July 26th, '61; Caldwell co.
Jason Morton, 5th Sergeant, e. July 26th, '61; Caldwell co.; dg. April, '62.
John E. Grier, 1st Corporal, e. July 26th, '61; Caldwell co.; p. Sergeant April, '62.
J. C. Grier, 2d Corporal, e. July 26th, '61; Caldwell co.; p. Sergeant, and 2nd Lieut. September, '62.
James Barnes, 3d Corporal, e. July 26th, '61; Caldwell co.; w. July 1st, '63, at Gettysburg.
Seth T. Dula, 4th Corporal, e. July 26th, '61; Caldwell co.; w. at Gettysburg, and k. in Virginia.

PRIVATES.

Amos, Franklin, e. July 26th, '61; Caldwell co.; dg. April, '62.
Anderson, James, e. July 26th, '61; Caldwell co.; dg. September, '62.
Angeley, J. B., e. March 15th, '62; Caldwell co.

Blair, R. M., e. October 21st, '61; Caldwell co.; p. Corporal in '63; w. at Gettysburg July 1st, '63; k. at Bristoe Station.

Bradshaw, James J., e. August 30th, '61; Caldwell co.; w. July 3d, '63, at Gettysburg.

Bradshaw, A. L., e. March 15th, '62; Caldwell co.; w. July 1st, '63, at Gettysburg.

Bush, J. A., e. November 30th, '61; Caldwell co.; p. Corporal April, '62; Sergeant September, '62; w. July 1st, '63, at Gettysburg.

Bush, Wyatt, e. November 30th, '61; Caldwell co.; d. August, '62.

Bush, N. M., e. March 15th, '62; Caldwell co.

Barlow, J. O. C., e. March 15th, '61; Caldwell co.; d. June 27th, '62, of w. received near Richmond.

Barlow, Thomas M, e. March 15th, '62; Caldwell co.

Barlow, Calvin, e. March 15th, '62; Caldwell co.; dg August, '62.

Barnett, Walter, e. March 15th, '62; Caldwell co.

Barnett, L. B., e. March 15th, '62; Caldwell co.; m. July 3d, '63, at Gettysburg.

Borders, Martin, e. July 8th, '62; Caldwell co.; d. January, '63.

Brown, Alvy, e. January 12th, '63; Caldwell co.

Barlow, W. L., e. March 15th, '64; Caldwell co.

Bently, Caleb, e. Oct. 28th, '64; Wake co.

Cobb, J. A., e. Dec. 14th, '63; Caldwell co.

Collins, William, e. January 12th, '63; Caldwell co.; w. and pr. at Malvern Hill; k. July 1st, '63.

Chaney, B. W., e. August 30th, '61; Virginia.

Caison, A. S., e. March 15th, '61; Caldwell co.; m. July 3d, '63, at Gettysburg.

Cruise, Robert, e. March 15th, '61; Caldwell co.; w. July 1st, '63, at Gettysburg.

Chandler, T. C., e. January 12th, '63; Caldwell co.; w. at Gettysburg.

Chandler, Henry, e. March 15th, '62; Caldwell co.

Cloer, G. W., e. March 15th, '62; Caldwell co.

Downs, Clinton, e. July 26th, '61; Caldwell co.: d. September, '62.

Downs, James S., e. October 26th, '61; Caldwell co.

Deal, William, e. July 26th, '61; Caldwell co.

Dickson, M. B., e. September 23d, '62; Rowan co.

Daniel, James, e. June 12th, '63; Caldwell co.

Dinkens, J. A., e. July 26th, '61; Caldwell co.; dg. April, '62.

Dula, S. P., e. November 30th, '61; Caldwell co.; p. 1st Lieut. April 14th, '62.

Dumas, J. F., e. Nov. 3d, '63; Caldwell co.

Dula, J. S., e. April 1st, '64; Caldwell co.

Earp, Wm., e. September 23d, '63; Wilkes co.; w. July 1st, '63, at Gettysburg.

Felts, John, e. July 26th, '61; Caldwell co.; d. April, '62.

Felts, Aaron, e. July 26th, '61; Caldwell co.; w. July 1st, '63, at Gettysburg.

Fincannon, Isaac, e. July 26th, '61; Caldwell co.; d. April, '62.

Friddle, John, e. January 5th, '62; Caldwell co.; k. July 3rd, '63, at Gettysburg.

Furgerson, John, e. January 12th, '63; Caldwell co.

Gibson, Harrison, e. July 26th, '61; Caldwell co.; w. July 1st, '63, at Gettysburg.

Gibson, Peyton, e. July 26th, '61; Caldwell co.

Gibson, W. M., e. March 15th, '62; Caldwell co.; w. July 1st, '63, at Gettysburg.

Green, John, e. August 30th, '61; Caldwell co.; dg. Oct., '61.

Green, William, e. March 15th, '62; Caldwell co.

Gilbert, V., e. March 15th, '62; Caldwell co.

Hartley, T. H., e. July 26th, '61; Caldwell co.; w. July 1st, '63, at Gettysburg.

Holder, Hezekiah, e. July 26th, '61; Caldwell co.; w. July 1st, '63, at Gettysburg.

Hall, A. J., e. July 26th, '61; Caldwell co.; w. July 3d, '63, at Gettysburg.

Hilton, Joel, e. July 26th, '61; Caldwell co.; dg. Feb., '62.

Hawn, Henry, e. July 26th, '61; Caldwell co.; dg. Feb., '62.

Holder, J. C., e. July 8th, '62; Caldwell co.; w. July 1st, '63, at Gettysburg.

Hood, S. H., e. March 15th, '62; Caldwell co.
Holden, Henry, e. March 15th, '62; Caldwell co.; w. July 1st, '63, at Gettysburg.
Holden, Abraham, e. May 15th, '63; Caldwell co.; w. July 1st, '63, at Gettysburg.
Holden, John, e. March, '63; Caldwell co.; d. April, '64.
Holden, James, e. March, '63; Caldwell co.; d. April, '64.
Hamby, William, e. September 20th, '62; Caldwell co.
Harris, Goodwin, e. January 12th, '63; Caldwell co.
Hendren, E. H., e. September 23rd, '62; Wilkes co.; w. July 1st, '63, at Gettysburg.
Hart, J. C., e. July 8th, '63; Caldwell co.
Hubbard, Isham, e. September 23rd, '62; Wilkes co.
Houk, J. M., e. November 30th, '61; Caldwell co.; w. July 1st, '63, at Gettysburg.
Hart, W. S., e. March 1st, '64; Caldwell co.
Houston, G. W., e. March 15th, '62; Caldwell co ; dg. April, '63.
Justice, T. N., e. November 30th, '61; Caldwell co.
Justice, J. M., e. September 19th, '61; Caldwell co.; tr.
Jones, W. L., e. March 1st, '63; Caldwell co.; tr. from 22nd Regiment; d. of w. received July 1st, '63, at Gettysburg.
Johnson, L. P., e. September 23rd, '62; Caldwell co.; d. of w. received July 1st, '63, at Gettysburg.
Jopling, B. W., e. July 26th, '61; Caldwell co.; d. September, '62.
Keller, Henry, e. May 28th, '64; Caldwell co.
Lewis, Isaac, e. Sept. 3d, '63; Caldwell co.
Lee, Clinton, e. July 26th, '61; Caldwell co.; dg. April, '62.
Long, J. M. e. July 26th, '61; Caldwell co.; d. July, '62.
Lowdermilk, W. D., e. March 15th, '62; Caldwell co.; d.
Laney, Emdempy, e. March 15th, '62; Caldwell co.
Laney, Robert, e. March 15th, '62; Caldwell co.; dg. May, '63, for w. received July 1st, '62, at Malvern Hill.
Laney, J. S., e. March, 15th, '62; Caldwell co.; pr. in '64.
Laney, Levi, e. March 15th, '62; Caldwell co.; k. July 1st, '63, at Gettysburg.
Laney, Peter, e. March 15th, '62; Caldwell co.; pr. in '64.
Laney, James C., e. July 18th, '62; Caldwell co.; w. July 1st, '63, at Gettysburg; pr. in '64.
Laxton, Levi, e. January 12th, '63; Caldwell co.; w. July 1st, '63, at Gettysburg.
Lafeness, Harvey, e. March 15th, '62; Caldwell co.; pr. in '64.
McPherson, William, e. September 23rd, '62; Chatham co.; dt.
McRay, Clinton, e. July 26th, '61; Chatham co.
Minish, Ephraim, e. July 26th, '61; Chatham co,; dg. May, '62.
Matney, Thomas W., e. July 26th, '61; Chatham co.
Massey, E. F., e. July 26th, '61; Chatham co.; d. April, '62.
Maberry, J. S., e. July 26th, '61; Chatham co.; d. Sept., '62.
Minston, Hawson, e. July 26th, '61; Chatham co.; d. July, '61.
Morgan, Noah, e. October 26th, '61; Chatham co.; d. April, '62.
Martin, A. D., e. June 21st, '61; Chatham co.; w. at Gettysburg.
Minton, Thomas, e. April 14th, '62; Chatham co.; tr. from Com. E April, '62.
Matney, J. W., e. March 25th, '62; Chatham co.; k. July, '63, at Gettysburg.
McRay, H. G., e. April 30th, '62; Chatham co.; c. at Gettysburg.
McReary, F. M., e. July 26th, '61; Caldwell co.; p. Corporal October, '63.
McRay, John, e. January 12th, '63; Chatham co.
Marley, B. W., e. September 23d, '62; Wilkes co.; w. July 1st, '63, at Gettysburg.
McGarie, A. J., e. September 23d, '62; Wilkes co.
Minton, William, e. March 15th, '62; Caldwell co.; d. April, '63.
McAdams, J. D., e. Nov. 30th, '61; Tennessee; tr. to the 7th Tennessee Regiment.
McLeod, John, e. Oct. 8th, '64; Caldwell co.
Oxford, W. D., e. Aug. 2d, '63; Caldwell co.

Prestwood, M. L., e. November 30th, '61; Caldwell co.
Prestwood, Fashions, e. March 15th, '62; Caldwell co.; k. July 3d, '63, at Gettysburg.
Peircy, W. W., e. March 15th, '62; Wilkes co.; w. July 1st, '63, at Gettysburg.
Pennell, Thomas, e. March 15th, '62; Caldwell co.; m. at Falling Waters.
Phillips, James, e. September 23d, '62; Wilkes co.; d. Jan., '63.
Pilkenton, Wesley, e. Sept. 21st, '62; Wilkes co.
Pilkenton, Murphy, e. September 23d, '62; Wilkes co.
Porlier, J. W., e. September 23d, '62; Wilkes co.
Porlier, G. W., e. September 23rd, '62; Wilkes co.
Payne, William, e. October 10th, '61; Caldwell co.; tr. to Com. F, October, '62.
Reid, W. T., e. July 26th, '61; Caldwell co.
Roberts, William, e. July 26th, '61; Caldwell co.; d. December, '61.
Roberts, D. H., e. July 26th, '61; Caldwell co.
Robinson, James, e. July 8th, '62; Caldwell co.; w. July 1st, '63, at Gettysburg.
Robinson, A. G., e. July 8th, '62; Caldwell co.
Raily, J. W., e. November 30th, '61; Caldwell co.
Simmons, James, e. July 26th, '61; Caldwell co.
Summerow, George, e. July 26th, '61; Caldwell co.; w. July 3d, '63, at Gettysburg.
Summerow, J. C., e. July 26th, '61; Caldwell co.; w. July 1st, '63, at Gettysburg.
Smith, Ephraim, e. July 26th, '61; Caldwell co.; dg. September, '63.
Small, Jesse, e. July 26th, '61; Caldwell co.; dg. September, '63.
Small, Kelly, e. July 26th, '61; Caldwell co.
Sell, George, e. July 26th, '61; Caldwell co.
Sanders, Noah, e. July 26th, '61; Caldwell co.; d. April, '62.
Shell, T. H., e. July 26th, '61; Caldwell co.; dg. April, '62.
Summerow, Peter, e. March 15th, '62; Caldwell co.; w. July 3d, '63, at Gettysburg.
Stallings, John, e. November 7th, '61; Greene co.; w. at Malvern Hill.
Stallings, Urias, e. March 15th, '62; Caldwell co.; m. July 3d, '63, at Gettysburg.
Simmons, Thomas, e. March 15th, '62; Caldwell co.; m. at Falling Waters.
Simmons, J. M., e. March 15th, '62; Caldwell co.; w. July 1st, '63, at Gettysburg.
Siler, William, e. September 23d, '62; Chatham co.; m. July 3d, '63, at Gettysburg.
Setser, Eli, e. September 23d, '62; Caldwell co.; m. at Falling Waters.
Sudderth, G. H., e. March 15th, '62; Caldwell co.; k. July 1st, '63, at Gettysburg.
Sudderth, J. G. G., e. March 15th, '62; Caldwell co.; p. 2d Lieut. October 14th, '61.
Seigle, Noah, e. September 23d, '62; Catawba co.
Segraves, Rufus, e. Aug. 30th, '61; Wake co.; dg.
Smith, John, e. Dec. 21st, '63.
Snell, G. W., e. July 9th, '62; Caldwell co.
Smith, H. W., e. Nov. 7th, '61; Caldwell co.
Stafford, Alfred, e. March 15th, '62; Caldwell co.; dg.
Talbert, Alex., e. Feb. 1st, '64; Caldwell co.; pr. in '64.
Teague, J. A., e. Dec. 20th, '61; Caldwell co.
Teague, Washington, e. March 15th, '62; Caldwell co.; d. Aug., '62.
Talbert, Giles, e. July 26th, '61; Caldwell co.; dg. April, '62.
Talbert, John, e. July 26th, '61; Caldwell co.; w. at Malvern Hill and Gettysburg.
Taylor, Milas, e. July 26th, '61; Caldwell co.; m. July 3d, '63, at Gettysburg.
Turnmire, John, e. March 15th, '62; Caldwell co.; d. October, '63.
White, Ambrose, e. July 26th, '61; Caldwell co.; m. at Falling Waters.
White, J. T., e. July 26th, '61; Caldwell co.; m. at Falling Waters.
White, J. E., e. July 26th, '61; Caldwell co.; d. April, '62.
Wilson, T. A., e. July 26th, '61; Caldwell co.
Wilson, Joseph, e. July 26th, '61; Caldwell co.; k. July 1st, '63, at Gettysburg.
Watts, Manly, e. July 26th, '61; Caldwell co.; m. at Falling Waters.
Watson, T. H., e. July 26th, '61; Caldwell co.; w. at Gettysburg.

West, Hiram, e. July 26th, '61; Caldwell co.; dg. April 1st, '62.
West, James A., e. July 26th, '61; Caldwell co.; k. July 1st, '63, at Malvern Hill.
Watson, J. J., e. July 8th, '63; Caldwell co.
Watson, D. W., e. Jan. 12th, '63; Caldwell co.
Wakefield, J. M., e. March 15th, '62; Caldwell co.; w. at Gettysburg.
Wilson, William, e. March 15th, '62; Caldwell co.; w. at Gettysburg.
White, J. T., e. Nov. 30th '63; Caldwell co.
White, Ambrose, e. Oct. 25th, '64; w. in '64.
Zimmerman, ——, e. March 15th, '62; Caldwell co.; d. July, '63.

COMPANY K.

OFFICERS.

James C. Calloway, Captain, cm. July 1st, '61; Anson co.
Thomas Lilly, Captain, cm. ———; Anson co.

J. S. Kendall, 1st Lieut., cm. July 1st, '61; Anson co.
Thomas Lilly, 1st Lieut., cm. April 20th, '62; Anson co.
John A. Polk, 1st Lieut., cm. ———; Anson co.
John C. McLauchlin, 2d Lieut., cm. July 1st, '61; Cumberland co.
William C. Boggan, 2d Lieut., cm. July 1st, '61; Anson co.
William L. Ingram, 2d Lieut., cm. April 21st, '62; Anson co.
W. D. Webb, 2d Lieut., cm. ———; Anson co.
Jesse L. Henry, 2d Lieut., cm. August 9th, '63; Anson co.

NON-COMMISSIONED OFFICERS.

Benjamin McLauchlin, 1st Sergeant, e. July 1st, '61; Cumberland co.; d. April 14th, '62.
Jesse L. Henry, 2d Sergeant, e. July 1st, '61; Anson co.; p. 2d Lieut. Aug. 9th, '63.
Wm. L. Ingram, 3d Sergeant, e. July 1st, '61; Anson co.; p. 2d Lieutenant April 21st, '62.
Edmond J. DeBerry, 4th Sergeant, e. July 1st, '61; Anson co.; tr. to 52d Regiment May 1st, '62.
John Q. Neal, 5th Sergeant, e. July 1st, '61; Montgomery co.; d. April 1st, '62.
Thomas E. Knoots, 1st Corporal, e. July 1st, '61; Anson co.; d. April 2d, '62.
Edmond C. Braswell, 2d Corporal, e. July 1st, '61; Anson co.
Joel T. Gaddy, 3d Corporal, e. July 1st, '61; Anson co.; w. July 1st, '63, at Gettysburg.
Thomas Lilly, 4th Corporal, e. July 1st, '61; Anson co.; p. 1st Lieutenant April 20th, '62.
John A. Polk, Musician, e. July 1st, '61; Anson co.; p. 1st Sergeant April 31st, '62.
Calvin G. Boyd, Musician, e. July 1st, '61; Anson co.; d. Sept. 8th, '62.

PRIVATES.

Allen, Robert B., e. July 1st, '61; Anson co.
Atkinson, John, e. July 1st, '61; South Carolina.
Allen, G. W., e. May 10th, '61; Anson co.; w. July 1st, '63, at Gettysburg.

Allen, W. C., e. May 10th, '61; Anson co.; w. at Gettysburg.
Allen, Josiah, e. Dec. 2d, '61; Anson co.; tr. to 43d Regiment Sept. 20th, '63.
Boyett, J. M., e. May 10th, '62; Anson co.; dg. Sept., '62, for w.
Bahannon, J. A., e. July 1st, '61; Anson co.; dg. Sept. 24th, '62.
Bowman, William, e. May 20th, '62; Anson co.; k. July 1st, '63, at Gettysburg.
Bowman, George, e. May 10th, '62; Anson co.; k. June 25th, '62, near Seven Pines.
Bryley, T., e. Sept. 19th, '62; Anson co.; d. of w. received at Gettysburg.
Boyd, Calvin R., e. July 1st, '61; Montgomery co.
Briley, John, e. July 1st, '61; Anson co.; p. Sergeant April 21st, '62.
Braswell, Benjamin F., e. July 1st, '61; Anson co.
Braswell, Daniel, e. July 1st, '61; Anson co.; w. July 1st, '63, at Gettysburg.
Bowen, Thomas, e. July 1st, '61; Anson co.
Barber, Sidney, e. July 1st, '61; Anson co.; dg. September 24th, '62, and re-enlisted July, '63.
Boylan, William C., e. July 1st, '61; Anson co.; k. July 1st, '63, at Gettysburg.
Broadway, William H., e. July 1st, '61; Anson co.; p. Corporal and Sergeant; w. July 1st, '63, at Gettysburg.
Burns, Wm. H., e. July 1st, '61; Anson co.; d. Sept. 4th, '62.
Barber, Riley, e. August 21st, '62; Anson co.
Beachman, Jeremiah W., e. May 10th, '62; Anson co.
Benton, Seabron A., e. July 17th, '62; Anson co.; w. July 1st, '63, at Gettysburg.
Browman, Thomas, e. July 1st, '61; Anson co.
Braswell, C. E., e. July 1st, '61; Anson co.
Buner, Thomas, e. May 10th, '62; Anson co.; w. July 1st, '63, at Gettysburg.
Broadway, John W., e. December 2d, '61; Anson co.
Burns, Julius W., e. May 10th, '62; Anson co.
Burr, Jacob, e. October 17th, '62; Anson co.
Carpenter, Isaac, e. July 1st, '61; Anson co.; w. July 3d, '63, at Gettysburg.
Carpenter, Calvin, e. May 10th, '62; Anson co.
Carpenter, E. L., e. May 10th, '62; Anson co.
Carpenter, James C., e. May 10th, '62; Anson co.; w. July 1st, '63, at Gettysburg.
Caudle, A. A., e. May 10th, '62; Anson co.
Cripps, Nathan, e. July 1st, '61; Anson co.; w. accidentally.
Cox, Julius M., e. July 1st, '61; Anson co.; dg. September 24th, '62.
Curtis, Elijah T., e. July 1st, '61; Anson co.; dg. September 24th, '62.
Crowson, Henry H., e. July 1st, '61; Anson co.; p. Corporal; w. at Gettysburg.
Caudle, A. A., e. May 20th, '64; Anson.
Carpenter, Laban, e. Feb. 1st, '62; Anson co.; d. July 1st, '63.
Coendle, G. D., e. July 1st, '62; Anson co.; d. March 1st, '63.
Curlee, Joel, e. May 10th, '62; Anson co.; d. July 28th, '62.
Carles, Gaithing, e. May 10th, '62; Anson co.; d. October 26th, '62.
Dumas, H. E., e. Dec. 2d, '61; Anson co.; p. Corporal and Sergeant; w. July 1st, '63, at Gettysburg.
DeBerry, E. J., e. May 10th, '62; Anson co.; p. Sergeant.
Dabbs, W. H., e. Sept. 17th, '61; Anson co.; p. Corporal; w. at Malvern Hill and dg. in '62.
Dabbs, David L., e. July 1st, '61; Anson co.; p. Sergeant; k. at Gettysburg.
Davis, Edmund M., e. May 10th, '62; Anson co.
Edwards, F. M., e. May 10th, '62; Anson co.
Edwards, John F., e. May 10th, '62; Anson co.; w. at Gettysburg.
Eddings, John R., e. December 2d, '61; Anson co.; w. at Gettysburg.
Edwards, Allen R., e. July 1st, '61; Georgia; d. of w. received at Gettysburg.
Edward, F. M., e. Feb. 24th, '64; Anson co.
Edward, A. B., e. May 10th, '63; Anson co.; k. at Gettysburg.
Edwards, John O., e. May 10th, '62; Anson co.; k. July 3d, '63, at Gettysburg.

Flake, P. H., e. May 10th, '62; Anson co.; k. July 3d, '63, at Gettysburg.
Flake, John, e. July 1st, '61; Anson co.; k. July 1st, '63, at Gettysburg.
Flake, Elijah W., e. January 30th, '63; Anson co.; tr. from C. S. Navy; w. at Gettysburg.
Gadd, James E., e. July 1st, '61; Anson co.; w. at Gettysburg.
Gathing, S. F., e. July 1st, '61; Anson co.; w. June 25th, '62.
Gulledge, Thomas, e. July 1st, '61; Anson co.; w. March 14th, '62, at Barrington Ferry, and Gettysburg.
Gray, Francis M., e. May 10th, '62; Anson co.; dt. in Ordnance Department.
Griffen, S. R., e. July 1st, '61; Anson co.; k. at Gettysburg.
George, F. M., e. July 1st, '61; Anson co.; dg. May 6th, '62.
Gulledge, William, e. July 1st, '61; Anson co.
Gathing, Wm., e. May 10th, '62; Anson co.
Gulledge, Hill, e. April 1st, '64; Anson co.; pr. in '64.
Green, Thomas, e. July 1st, '61; Anson co.; k. July 3d, '63, at Gettysburg.
Gaddy, E. D., e. December 2d, '61; Anson co.; k. July 3d, '63, at Gettysburg.
Gaithing, T. C., e. May 10th, '62; Anson co.; k. July 1st, '63, at Gettysburg.
Hyatt, J. W., e. May 1st, '63; Anson co.; k. July 1st, '63, at Gettysburg.
Horn, Elisha J., e. May 10th, '62; Anson co.; d. Aug. 21st, '62.
Hiad, Benjamin, e. July 1st, '61; Anson co.; dt. to Hospital.
Hildreth, M., e. May 10th, '62; Anson co.; dg. May 6th, '62.
Horne, W. T., e. May 10th, '62; Anson co.; dg. in '62.
Hinnant, Robert A., e. July 1st, '61; Anson co.; w. July 1st, '63, at Gettysburg.
Huntley, Edmund P., e. July 1st, '61; Anson co.
Hunt, Benjamin, e. July 1st, '61; New Jersey.
Harrington, Washington, e. May 10th, '62; Anson co.
Henly, James T., e. December 2d, '61; Anson co.
Hildreth, Elijah, e. September 17th, '61; Anson co.; w. July 1st, '63, at Gettysburg.
Hildreth, Robert, e. May 10th, '62; Anson co.
Hutchinson, George, Anson co.
Horn, Louis E., e. May 10th, '62; Anson co.; w. July 1st, '63, at Gettysburg.
Howard, James T., e. May 10th, '62; Anson co.; w. July 3d, '63, at Gettysburg.
Ingram, Joseph, e. May 1st, '63; Anson co.
Ingram, William W., e. July 1st, '61; Anson co.; w. July 1st, '63, at Gettysburg.
Jarman, Charles H., e. July 1st, '61; Anson co.
Jarman, John R., e. July 1st, '61; Anson co.
Jarman, Thomas D., e. July 1st, '61; Anson co.; w. July 1st, '63, at Gettysburg.
Jarman, Willis H., e. September 17th, '61; Anson co.; w. July 3d, '63, at Gettysburg.
Johnson, Moreland, e. December 2d, '61; Anson co.; w. July 3d, '63, at Gettysburg.
Johnson, Stanmore, e. May 10th, '62; Anson co.; w. July 3d, '63, at Gettysburg.
Jarmon, Elijah C., e. July 1st, '61; Anson co.; d. July, '63, of w. received at Gettysburg.
Jarmon, John H., e. July 1st, '61; Anson co.; p. Sergeant May, '62; d. August 2d, '62.
James, Wm. H., e. March 1st, '64; Anson co.; pr. in '64.
Kendall, Samuel S., e. July 1st, '61; Stanly co.; w. July 1st, '63, at Gettysburg.
Knotts, Benjamin F., e. July 1st, '61; Anson co.; dg.
La-Haisc, Oliver, e. September 11th, '61; Anson co.; w. March 14th, '63, at Barrington's Ferry.
Liles, Daniel S., e. July 1st, '61; Anson co.; w. July 1st, '63, at Gettysburg.
Lee, Franklin, e. May 10th, '61; Anson co.
Lee, John F., e. May 10th, '61; Anson co.

Lee. Henry D., e. February 25th, '62; Anson co.; tr. from 43d Regiment March, '63.
Laird, William J., e. May 10th, '62; Anson co.; w. July 3d, '63, at Gettysburg.
Little, Asa, e. July 1st, '61; Anson co.
Lee, Joseph J., e. July 1st, '61; Anson co.; k. July 1st, '63, at Gettysburg.
Liles, J. G., e. May 10th, '62; Anson co.; k. July 3d, '63, at Gettysburg.
Livingston, J. C., e. May 10th, '62; Anson co.; d. Aug. 3d, '62.
McLauchlin, W. A., e. May 1st, '62; Anson co.; d. July 9th, '62.
Moore, Wilson, e. May 10th, '62; Anson co.; d. August 2d, '63.
McGougan, P. N., e. May 10th, '62; Anson co.; d. Aug. 12th, '62, of w. received at Malvern Hill.
May, Thomas, e. July 1st, '61; Anson co.
McDiarmid, Martin, e. July 1st, '61; Anson co.; w. July 1st, '63, at Gettysburg.
Meachum, J. H., e. May 10th, '62; Anson co.
Myers, William A., e. May 10th, '62; Anson co.; w. July 3d, '63, at Gettysburg.
McRea, Montfort S., e. July 1st, '61; Montgomery co.; p. Sergeant and Sergeant Major; d. September, '63, of w. received at Fredericksburg.
Neal, John Q., e. July 1st, '61; Anson co.; p. Sergeant.
Nonee, Isham, e. August 3d, '61; Anson co.
Pinkston, Hugh D., e. July 1st, '61; Anson co.
Phillips, Terrell F., e. May 10th, '61; Anson co.
Phillips, Washington F., e. May 10th, '61; Anson co.; w. July 1st, '63, at Gettysburg.
Popli ., John, e. May 10th, '61; Anson co.; w. July 1st, '63, at Gettysburg.
Pope, John M., e. July 1st, '61; Anson co.; w. July 1st, '63, at Gettysburg.
Puvet, Elijah B., e. July 1st, '61; Anson co.; dg. November 21st, '62.
Polk, John A., e. July 1st, '61; Anson co.; p 2d Lieutenant.
Pinkston, Hugh, e. July 1st, '61; Anson co.
Pope, John M., e. July 1st, '61; Anson co.
Pool, Wm. R., e. July 8th, '62; Wake co.
Polk, L. L., e. May 10th, '61; Anson co.; p. Sergt. Major June, '62; p. 2d Lieut. of 43d Regiment.
Parrish, H. H., e. July 10th, '62; Anson co.
Rickett, Alexander, e. July 1st, '61; Anson co.; d. April 9th, '62.
Sanders, William Freeman, e. February 20th, '63; Anson co.
Smart, William D., e. July 1st, '61; Anson co.
Short, Jesse, e. July 1st, '61; Anson co.
Short, John B., e. July 1st, '61; Anson co.; w. at Rolles' Mill and Gettysburg.
Short, Samuel, e. July 1st, '61; Anson co.; w. November 9th, '62, at Rolles' Mill.
Short, William, e. September 26th, '62; Anson co.; w. July 1st, '63, at Gettysburg.
Spencer, Samuel T., e. ——; Anson co.; dg. January 4th, '62.
Short, Alexander, e. July 1st, '61; Anson co.
Sellers, Roland R., e. July 1st, '61; Anson co.; dg. June 5th, '62.
Sanders, Samuel T., e. July 1st, '61; Anson co.; d. April 9th, '62.
Smith, Wellington D., e. July 1st, '61; Anson co.; dg. May 25th, '62.
Short, Jesse D., e. July 1st, '61; Anson co.
Sanders, J. C., e. May 10th '62; Anson co.; d. August 10th, '62.
Sinclair, G. W., e. May 10th, '62; Anson co.; d. August 14th, '62.
Sullivan, J. D., e. July 1st, '61; Anson co.; pr. and d. August, '63.
Scarborough, J. B., e. May 10th, '63; Anson co.; k. at Gettysburg.
Smith, W. H., e. July 1st, '61; Anson co.; p. Sergeant and d. of w. received at Gettysburg.
Tyson, Albert, e. May 10th, '63; Anson co.; k. July 1st, '63, at Gettysburg.
Taylor, S. H., e. May 10th, '62; Anson co.
Teal, William H., e. July 1st, '61; Anson co.; d. April 9th, '63.
Teal, James A., e. May 10th, '62; Anson co.; w. July 1st, '63, at Gettysburg.

Teal, Joseph P., e. May 10th, '62; Anson co.; w. at Rolles' Mill Nov., '62, and Gettysburg July, '63.
Teal, Miles W., e. May 10th, '62; Anson co.; w. at Gettysburg.
Thomas, Calvin, e. May 10th, '62; Anson co.; w. at Gettysburg.
Thomas, J. W., e. May 10th, '62; Anson co.; w. at Gettysburg.
Tyson, J. A., e. May 10th, '62; Anson co.; w. at Gettysburg.
Teal, William T., e. July 1st, '61; Anson co.; d. Nov. 2d, '62.
Turnage, Luke, e. July 1st, '61; Anson co.; d. March 31st, '62.
Thomas, G. P., e. July 1st, '61; Anson co.
Wadsworth, J. F., e. May 10th, '62; Anson co.; w. July 1st, '63, at Gettysburg.
Willoughby, Rayford, e. July 1st, '61; Anson co.
Webb, W. D., e. December 2d, '61; Anson co.; p. 2d Lieutenant.
Woodburn, J. D., e. December 2d, '61; Anson co.
Wiggs, John, e. July 1st, '61; Anson co.; k. July 1st, '63, at Gettysburg.
Willoughby, Hiram, e. July 1st, '61; Anson co.; d. June 29th, '62, of w. received at Seven Pines.
Winfield, J. P., e. May 10th, '63; Anson co.; w. June 25th, '62, near Seven Pines and k. Nov. 2d, '63, at Rolles' Mill.
Woodbury, J. D., c. Dec. 2d, '61; Anson co.

TWENTY-SIXTH REGIMENT.

By ASSISTANT SURGEON GEORGE C. UNDERWOOD.

"Vixere fortes ante agamemnona multi; sed omnes illacrimabiles. urgentur ignotique longa nocte, carent quia vate sacro. Paulum sepultæ distat inertiæ celata virtus."

CAMP OF INSTRUCTION.

The regiment was mobilized at the Camp of Instruction at "Crab Tree," about three miles from Raleigh, N. C. At this Camp, during the months of July and August, 1861, were assembled ten companies from the counties of Alamance, Anson, Ashe, Caldwell, Chatham, Moore, Randolph, Union, Wake, and Wilkes. These companies were organized before leaving home, and on arrival at Camp of Instruction, reported as follows:

1.—Jeff Davis Mountaineers, Ashe County; Captain, Andrew N. McMillan; First Lieutenant, George R. Reeves; Second Lieutenant, Jesse A. Reeves; Junior Second Lieutenant, James Porter.

2.—Waxhaw Jackson Guards, Union County; Captain, J. J. C. Steele; First Lieutenant, William Wilson; Second Lieutenant, Taylor G. Cureton; Junior Second Lieutenant, John W. Richardson.

3.—Wilkes Volunteers, Wilkes County; Captain Abner R. Carmichael; First Lieutenant, Augustus H. Horton; Second Lieutenant, Phineas Horton; Junior Second Lieutenant, William W. Hampton.

4.—Wake Guards, Wake County; Captain, Oscar R. Rand; First Lieutenant, James B. Jordan; Second Lieutenant, James T. Adams; Junior Second Lieutenant, James W. Vinson.

5.—Independent Guards, Chatham County; Captain, W.

S. Webster; First Lieutenant, William J. Headen; Second Lieutenant, Bryant C. Dunlap; Junior Second Lieutenant, S. W. Brewer.

6.—Hibriten Guards, Caldwell County; Captain, Nathaniel P. Rankin; First Lieutenant, Joseph R. Ballew; Second Lieutenant, John B. Holloway; Junior Second Lieutenant, Alfred T. Stewart.

7.—Chatham Boys, Chatham County; Captain, William S. McLean; First Lieutenant, John E. Matthews; Second Lieutenant, George C. Underwood; Junior Second Lieutenant, Henry C. Albright.

8.—Moore Independents, Moore County; Captain, William P. Martin; First Lieutenant, Clement Dowd; Second Lieutenant, James D. McIver; Junior Second Lieutenant, Robert W. Goldston.

9.—Caldwell Guards, Caldwell County; Captain, Wilson S. White; First Lieutenant, John Carson; Second Lieutenant, John T. Jones; Junior Second Lieutenant, Milton P. Blair.

10.—Pee Dee Wild Cats, Anson County; Captain, James C. Carraway; First Lieutenant, James S. Kendall; Second Lieutenant, John C. McLauchlin; Junior Second Lieutenant, William C. Boggan.

The commandant of the Camp of Instruction at Crab Tree was Major Harry King Burgwyn, Jr., not twenty-one years of age, who had graduated at the Virginia Military Institute in May previous.

The Adjutant of the Camp was Oliver Cromwell Petway, also a cadet at the Virginia Military Academy in 1860-1861, subsequently Lieutenant-Colonel of the Thirty-fifth North Carolina Regiment, and killed at Malvern Hill 1 July, 1862.

Of this young commandant, Corporal John R. Lane, Company G, subsequently rising by his military talents to the Colonelcy of the regiment, gives his first impressions as follows: "We took the train at Company Shops (now Burlington) for Raleigh; arriving at this place, the company marched out to Camp Crab Tree, a Camp of Instruction, and were assigned our position in camp a little after dark. On the next morning when we awoke, we saw the sentinels at their posts and realized that we were indeed in the war. Immediately after roll call—but there was no roll call in our company—Major H. K. Burgwyn, commander of the Camp of Instruction, sent down to Captain W. S. McLean, demanding the reason for his failure to report his company.

Before the excitement occasioned by his message had subsided among the commissioned officers, an order came for a corporal and two men to report at once at headquarters. Captain McLean selected Corporal Lane, his lowest subaltern officer, and two of the most soldierly-looking men, S. S. Carter and W. G. Carter, to report to Major Burgwyn.

Accordingly, these three worthies appeared before the commandant, wondering whether they were going to be promoted, hanged or shot. This was our first sight of the commanding officer, who appeared though young, to be a youth of authority, beautiful and handsome; the flash of his eye and the quickness of his movements betokened his bravery. At first sight I both feared and admired him. He gave us the following order: "Corporal, take these men and thoroughly police this Camp; don't leave a watermelon rind or anything filthy in Camp."

This cheering order completely knocked the starch out of our shirts and helped greatly to settle us down to a soldier's life. The cleanliness of the camp was reported by the officer of the day as being perfect. You may be sure our officers reported the company promptly after that.

REGIMENTAL ORGANIZATION (AUGUST 27, 1861).

The companies composing the regiment were from the central and western counties of the State; counties which had opposed secession until the Proclamation of President Lincoln (April 15, 1861) calling upon Governor Ellis to furnish North Carolina's quota of seventy-five thousand volunteers to coerce the seceding Southern States.

After being drilled and otherwise disciplined, these ten companies were organized into a regiment designated as the

Twenty-sixth North Carolina Troops (Infantry) and the companies took rank as follows:

CAPTAIN McMILLAN'S COMPANY, from Ashe County, as Company A.
CAPTAIN STEELE'S COMPANY, from Union County, as Company B.
CAPTAIN CARMICHAEL'S COMPANY, from Wilkes County, as Company C.
CAPTAIN RAND'S COMPANY, from Wake County, as Company D.
CAPTAIN WEBSTER'S COMPANY, from Chatham County, as Company E.
CAPTAIN RANKIN'S COMPANY, from Caldwell County, as Company F.
CAPTAIN McLEAN'S COMPANY, from Chatham County, as Company G.
CAPTAIN MARTIN'S COMPANY, from Moore County, as Company H.
CAPTAIN WHITE'S COMPANY, from Caldwell County, as Company I.
CAPTAIN CARRAWAY'S COMPANY, from Anson County, as Company K.

The company officers completed the regimental organization by electing as Colonel, Captain Zebulon B. Vance, then Captain of the "Rough and Ready Guards" from Buncombe County, in the Fourteenth North Carolina Troops; as Lieutenant-Colonel, Major Harry K. Burgwyn, Jr., commandant of the camp; and as Major, Captain Abner B. Carmichael, of Company C.

Colonel Vance subsequently appointed First Lieutenant James B. Jordan, of Company D, Adjutant; Sergeant Joseph J. Young, of Company D, Quartermaster; Lieutenant Robert Goldston, of Company H, Commissary, who died at Carolina City October, 1861; Dr. Thomas J. Boykin, of Sampson County, Surgeon; and Private Daniel M. Shaw, Company H, Assistant Surgeon.

Rev. Robert H. Marsh, of Chatham County, since so widely

known as an eloquent preacher of the Baptist persuasion, was appointed Chaplain. The commissions of the officers bore date 27 August, 1861. First Lieutenant A. H. Horton, of Company C, was promoted Captain vice Carmichael, elected Major. The non-commissioned staff were:

L. L. Polk, Sergeant-Major, of Company K.
Benjamin Hind, Hospital Steward, of Company K.
E. H. Hornaday, Ordnance Sergeant, of Company E.
Jesse Ferguson, Commissary Sergeant, of Company C.
Abram J. Lane, Quartermaster Sergeant, of Company G.

ENCAMPMENT ON BOGUE ISLAND.

Promptly on its organization the regiment was ordered to the defence of Fort Macon, on Bogue Island. Leaving Raleigh on the 2d of September, 1861, under command of Lieutenant-Colonel Burgwyn, (Colonel Vance not having as yet reported for duty), the regiment, halting a few days at Morehead City, took up its permanent camp near Fort Macon—at which place Colonel Vance assumed command. The months of September, October and November, 1861, were passed at this place. The time was occupied in guard duties, drilling and preparing for the arduous duties that lay before them.

Occasionally, upon rumor that the enemy were landing, the long roll would be sounded, and the regiment drawn up in line. There was great sickness among the soldiers. An endemic of measles and fever prevailed. A hospital was established at Carolina City on the mainland, three miles west of Morehead City—Commissary Goldston, Assistant Surgeon Shaw, Lieutenant John E. Matthews and many privates died in a short while. Nine men from one Company died in a week. Supplies had to be brought across the Sound, and the water being shallow, the men had to wade quite a distance to get to the vessels bringing the rations.

The regimental officers were incessant in their attentions to their men, showing them every kindness, providing every comfort possible, and became much endeared to those under their authority. When time came to go into winter quarters, the regiment was moved to the mainland and camped midway between Morehead and Carolina Cities. While in this camp, Captain McLean, of Company G, was appointed Acting Assistant Surgeon, and Corporal John R. Lane elected Captain of the Company.

The winter of 1861-1862 was passed in unremitting drill and under strict measures of discipline, which got the regiment into fine condition for the opening campaign; and here they acquired a reputation for efficiency in drill and obedience to orders which they retained with increasing credit until the final surrender

In October, 1861, General D. H. Hill was appointed to the command of the District of Pamlico, to be succeeded in November by Brigadier-General L. O'B. Branch. After the fall of Roanoke Island (10 February, 1862) and in view of the threatened attack on New Bern by General Burnside, the regiment was ordered up the railroad within three miles of New Bern, and there went into bivouac and assigned to Branch's command, which as then constituted, was composed of the Seventh, Twenty-sixth, Twenty-seventh, Thirty-third, Thirty-fifth and Thirty-seventh North Carolina Regiments, Infantry, and Latham's and Brem's Batteries of artillery, Colonel Spruill's Second Cavalry (Nineteenth North Carolina), a battalion of militia under Colonel H. J. B. Clark, and some detached companies. Brigadier-General R. C. Gatlin, commanding the Department of North Carolina and coast defenses, headquarters at Goldsboro, was in supreme command.

BATTLE OF NEW BERN, N. C. 14 MARCH, 1862.

General Ambrose E. Burnside flushed with his captures of Fort Hatteras (29 August, 1861) and Roanoke Island (10 February, 1862) was now about to attempt still greater movements on the military chess board, and on 11 March, 1862, he embarked the brigades of Foster, Reno and Parke and accompanying artillery, at Roanoke Island and reached Slocum's Creek where it empties into the Neuse river, some sixteen miles from New Bern, on the evening of the 12th. Early next morning, after shelling the country around, General Burnside disembarked his command, and ordered Foster's

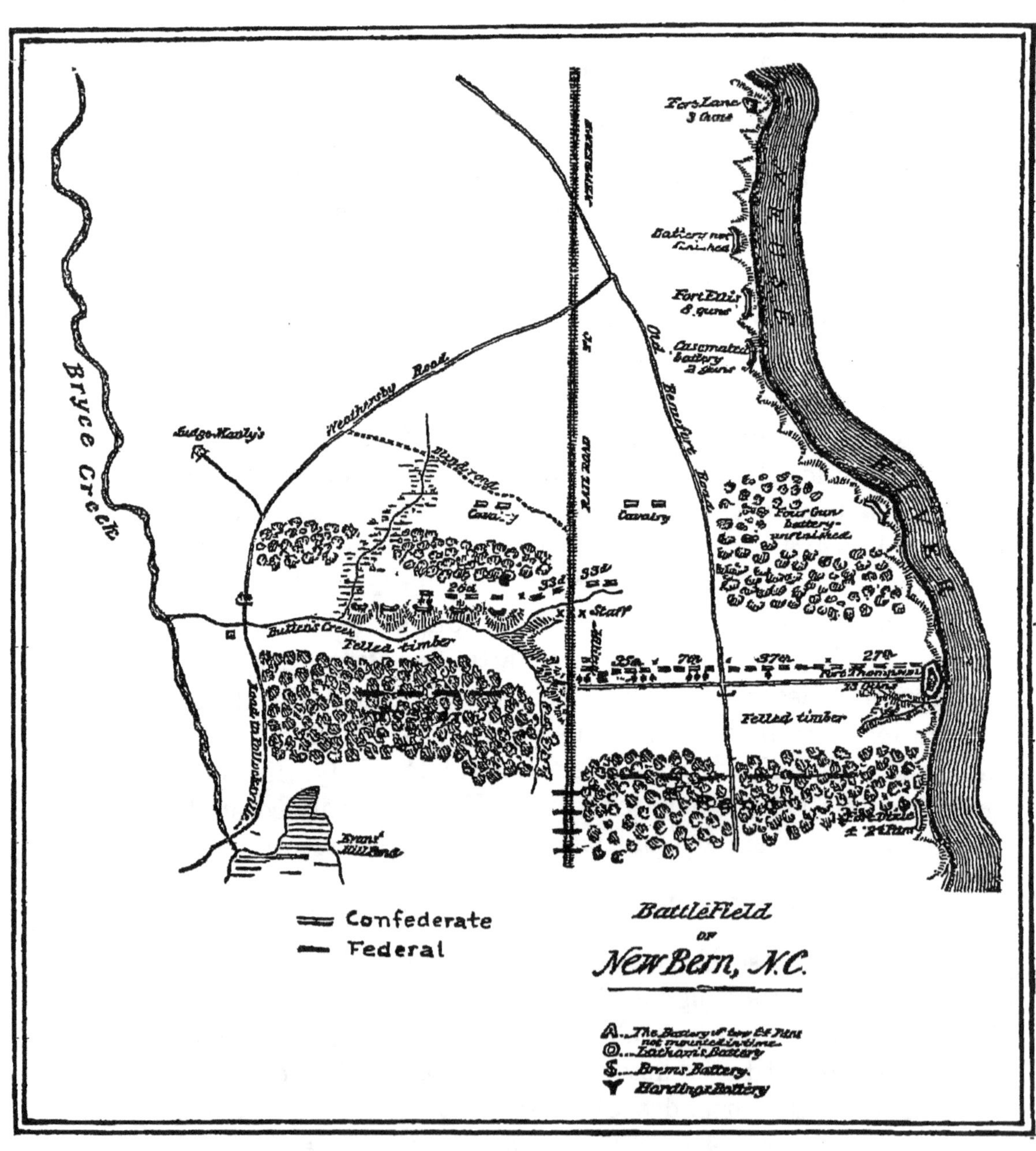
Bryce Creek
Weathersby Road
Rail Road
Old Beaufort Road
Cavalry
Felled timber
26th
33d
Staff
35th
7th
37th
27th
Fort Thompson
Fort Ellis 8 guns
Four Gun battery unfinished
Confederate
Federal
Battlefield of New Bern, N.C.

Brigade to advance up the county road, and attack our front and left; Reno's Brigade to march up the railroad with orders to turn our right; and Parke's Brigade to follow along the county road at convenient distance as a support either to Foster or Reno as there might be need.

General Burnside's advance appears to have met no opposition; the Croatan breastworks above Otter Creek he found abandoned, and at night his entire command bivouacked in easy striking distance of the Confederate lines of defence, which we will now describe.

About five miles below New Bern on the right bank of the Neuse River the Confederates had constructed a strong fort, called "Fort Thompson," manned by thirteen siege guns of good size, supported by ten field pieces, with three navy 32-pounders on its rear face.

From the fort in a straight line to the railroad leading from New Bern to Morehead City, was the main line of defense, consisting of a strong breastwork about one and one-quarter miles in length.

Through the centre of these breastworks the Beaufort County road leading to New Bern passed, and intersected the railroad about two miles behind the works; thence crossed the Trent river on a wooden bridge about a mile and a half above New Bern. Where the breastworks met the railroad there was a brick kiln, and this proved to be the cause of all our woes in this battle. Instead of continuing the breastworks straight across the railroad into the swamp beyond, to make the line as short as possible after reaching the railroad, the line was thrown back about 150 yards to the banks of Bullen's Creek and thence, a series of small breastworks conforming to the features of the ground, ran off in the direction of a swamp. To guard this gap of 150 yards in which was this brick kiln plant, General Branch ordered the brick kiln to be loopholed; and the evening before the battle, had ordered down two 24-pounder guns which were being mounted when the party was fired into in the beginning of the action and the work was stopped never to be resumed. The timber in front of the breastworks had been felled for some 350 yards.

General Branch's disposition of his troops had to be made

with great rapidity, as the enemy left him no time for delay. At 4 p. m. on the 12th of March, General Branch was notified of the enemy's approach. He ordered Colonel Sinclair, of the Thirty-fifth North Carolina Regiment, to proceed to Fisher's landing, just above the mouth of Otter Creek, to resist any attempt of the enemy to land. Late in the night he ordered the Twenty-sixth North Carolina Regiment and Brem's Battery, Lieutenant-Colonel Burgwyn in command, to follow, Colonel Vance being temporarily in command of the Post of New Bern. These troops arrived to find the enemy had anticipated them by occupying this ground, so the two regiments fell back to take their places in the main line for the next day's battle.

General Branch divided his forces that were to defend the works on the left of the railroad, namely, between the railroad and Fort Thompson, into two wings to be commanded respectively by Colonel C. C. Lee, of the Thirty-seventh North Carolina Regiment, and Colonel Reuben P. Campbell, of the Seventh North Carolina Regiment. Colonel Lee's command embraced the troops between the fort and the county road, and was composed of the Twenty-seventh North Carolina Regiment and his own, the Thirty-seventh North Carolina Regiment; on the right of the county road reaching to the railroad constituted Colonel Campbell's command and was defended by his own regiment (the Seventh) ; the Thirty-fifth and Captain Whitehurst's independent company, and on the right next to the railroad was placed the battalion of militia under command of Colonel H. J. B. Clark. Two sections of Brem's and Latham's batteries of artillery were posted along this line between the county road and railroad, under Colonel Campbell's command.

Colonel Vance, of the Twenty-sixth North Carolina Regiment, was in command of all the defences on the right of the railroad, comprising a distance of one and a quarter miles. His own regiment, one or two detached companies and a section of Brem's artillery, were the only troops at his disposal for this important defense. His line ran along the bank of Bullen's Creek for about half a mile, until the creek emptied into a swamp; beyond this swamp his line was extended to the

Weathersby road leading to New Bern; and beyond this (on the right) was Bryce's Creek, a deep and impassable stream of about 75 yards wide, which empties into the Trent River. Shortly after the battle opened, the part of Governor Vance's line next to the railroad and under the immediate command of Major Carmichael, of the Twenty-sixth Regiment, was re-inforced, first by five companies of Colonel Avery's Regiment, the Thirty-third North Carolina, held in reserve; and as the battle progressed and more determined became the attempt of the enemy to carry this position, the other five companies of the Thirty-third Regiment, under the gallant Colonel Avery and Lieutenant-Colonel Robert F. Hoke, came to Major Carmichael's assistance. As will hereafter be seen, the enemy never succeeded in carrying the works on the right of the railroad.

During the day of the 13th, the enemy kept up a brisk shelling from their gun boats, now in the Neuse, and keeping abreast of their land forces; and by night had gotten his three brigades in position for the attack early the next morning. These were disposed as follows: General J. G. Foster formed his line across the county road parallel to the Confederate works, the Twenty-fifth and Twenty-fourth Massachusetts Regiments on the right, and the Twenty-seventh and Twenty-third Massachusetts on the left, supported by six navy howitzers and the howitzers of Captains Dayton and Bennett.

General Jesse L. Reno formed his brigade on the left of the railroad in the following order, viz., the Twenty-first Massachusetts, Ninth New Jersey and Fifty-first Pennsylvania Regiments. General Parke's Brigade was drawn up in line in the intermediate space between General's Foster and Reno, with orders to support whichever brigade needed it.

About 7:30 a. m. the battle was opened by a shot from a Parrott gun from Latham's battery under Lieutenant Woodbury Wheeler. This shot dispersed a squad of horsemen who seemed to be reconnoitering under cover of the woods. Immediately after this, the firing became general. General Foster's attacks on the main works in his front made but little, if any, impression; they were easily repulsed. Doubtless the enemy knew the weak points in the Confederate line of defense. Immediately on getting his men into line, General Reno ordered Lieutenant-Colonel W. S. Clark to charge with the right wing of his regiment, the Twenty-first Massachusetts, and take the brick kiln.

Colonel Clark says in his report: "At the moment of our arrival at the Cut, the enemy were busily engaged in removing ammunition from the cars which had just come down from New Bern with re-enforcements. At the first volley from Company C the enemy in great astonishment, fled from the road and trench to a ravine in the rear of the brick yard. General Reno ordered Color-bearer Bates to plant his flag upon the roof of a building within the enemy's intrenchments. General Reno, with Companies C, A, B, and H, of the right wing, dashed across the railroad up the steep bank and over the rifle trench on top into the brick yard. Here we were subjected to a most destructive cross fire from the enemy on both sides of the railroad and lost a large number of men in a very few minutes. The General supposing he had completely flanked the enemy's works, returned across the road to bring up the rest of his brigade; but just at this time a tremendous fire of musketry and artillery was opened from the redoubts hitherto unseen, which were nine in number, extending from the railroad more than a mile to the right into the forest.

"The General, now obliged to devote his attention to the enemy in front of his brigade, ordered the left wing of the Twenty-first Massachusetts not to cross the road, but to continue to fire upon the enemy in the first two redoubts. These troops consisted of the Thirty-third North Carolina and the Twenty-sixth North Carolina Regiments, and were the best armed and fought the most gallantly of any of the enemy's forces; their position was almost impregnable so long as their left flank resting on the railroad was defended. They kept up an incessant fire for three hours until their ammunition was exhausted, and the remainder of the rebel forces had retreated from that portion of their works lying between the river and the railroad."

Having quoted so freely from the Federal side, let us now see what was doing among the Confederates. It is seen, General Branch had but one brigade to oppose three—but six regiments to oppose thirteen. These thirteen Federal regiments were in full ranks. The Twenty-first Massachusetts, of which we have been speaking, took into the battle 743 men. When Colonel Campbell was informed by Colonel Sinclair, "under much excitement," that the enemy had flanked him and were coming up the trenches which had been vacated by the militia, Colonel Campbell ordered Colonel Sinclair to leave the works and charge bayonets upon the advancing columns; this Colonel Sinclair failed to do, and left the field in confusion. Colonel Campbell then ordered Lieutenant-Colonel Haywood to have his men, the Seventh North Carolina Regiment, leave the works and charge the enemy. This was done in handsome style, and the enemy were driven over the breastworks and the guns of Brem's Battery that had fallen into their hands, were retaken. This charge was so impetuous that the enemy largely magnified the number of men that made it. Says Colonel Clark, of the Twenty-first Massachusetts Regiment, resuming our quotation from his report of the battle: "Having been ordered into the brick yard and left there with my colors and the four companies above mentioned, and finding it impossible to remain there without being cut to pieces, I was compelled either to charge upon Captain Brem's Battery of flying artillery or retreat. Accordingly, I formed my handful of men (about 200) in line, the right resting on the breastworks of the enemy, and commenced firing upon the men and horses of the first piece. Three men and two horses having fallen, I gave the order to charge bayonets and went to the first gun. Leaving this in the hands of Captain Walcott and Private John Dunn, of Company B, I proceeded to the second gun, about 300 paces from the brick yard. By this time the three regiments of the rebel infantry, who had retreated from the breastworks to a ravine in the rear when we entered the brick yard, seeing that we were so few and received no support, rallied and advanced on us. The Thirty-fifth and Thirty-seventh North Carolina, supported by the Seventh North Carolina, came upon us from the ravine in splendid style, with their muskets at the right shoulder and halted. Most fortunately, or rather providentially, for us, they remained undecided for a minute or two, and then resolved on a movement which saved us from destruction. Instead of giving us a volley at once, they first hesitated, and then charged upon us without firing. I instantly commanded my men to spring over the parapet and ditch in front, and to retreat to the railroad, keeping as close as possible to the ditch. On the railroad I found Colonel Rodman with the Fourth Rhode Island Regiment waiting for orders, and I urged him to advance at once and charge upon their flank, as I had done."

Up to this point in the battle, everything had gone on satisfactorily for the Confederates on the right of the railroad. General Reno's attacks had been met and repulsed handsomely. The Confederate line of defense on the right of the railroad as heretofore stated, consisted of rifle pits and detached intrenchments in the form of lunettes and redans along the bank of Bullen's Creek, and across the swamp to the Weathersby road, about one and one-quarter miles. A rifle pit near the railroad was occupied by Captain Oscar R. Rand, with his Company D, about 77 men; by Company A, 68 men, and by 25 men from Company G, all under command of Major A. B. Carmichael, of the Twenty-sixth Regiment. Quoting from Captain Rand's account of the battle, written shortly after his capture and addressed to Colonel Z. B. Vance:

"About 7:30 a. m. the battle commenced on the left and for a time, extending from Fort Thompson along the whole line of the breastworks to the railroad, the roar of cannon and musketry was incessant. Within a few minutes after the battle had commenced, the enemy made his appearance on the right of the railroad directly in front of us. About one regiment (the left wing of the Twenty-first Massachusetts) took position between the railroad and Bullen's Creek, sheltering themselves in the woods and behind the logs, while the main body consisting of several regiments advanced under cover of the woods down the opposite side of the creek, occupying the heights and extending himself along our right.

"When the advance of the enemy had reached nearly opposite Major Carmichael's position, he gave the order to fire, and sent a volley full into the head of the advancing column. The enemy replied immediately and from this time to the close of the action, the firing never ceased. At first, the enemy shooting very badly, their balls flying high above our heads and cutting the boughs from the tops of the trees in our rear, whereas our men, under direction of Major Carmichael and other officers, took deliberate aim, sending death into their ranks. As soon as we were fairly engaged with this part of the enemy, the other part which held position between the railroad and the creek came up from under their cover and attempted to cross the railroad with a view to flank the main intrenchments and cut our lines in two.

"No sooner was this attempted than it was discovered, and every gun ordered to bear upon them. One well directed volley scattered this force. Many a poor fellow fell here to rise no more, for they were well exposed.

"Just at this time, about half an hour after the battle had commenced, Colonel Avery, who had been held in reserve, arrived with the Thirty-third regiment. He with four companies entered the rifle pits occupied by us, while four other companies under Major Gaston Lewis, were ordered to occupy an advanced rifle pit nearest to the brick yard. This movement was attended with great danger, and was gallantly executed. Major Lewis had to advance a space of 150 yards over fallen timber; all the while exposed to the enemy's fire, and without being able to return it. He gained the position, however, and held it during the remainder of the action.

"The battle now raged furiously; the enemy throwing themselves along our right so as to gain the point from which he could fire directly into our trenches, and Colonel Avery, ably seconded by Major Carmichael, using every effort to prevent it. In this they were somewhat aided by the artillery and infantry, part of the Twenty-sixth Regiment and two companies of the Thirty-third Regiment, under Lieutenant-Colonel Hoke—on the right of us, only two or three companies of which, however, were within range. The intention of the enemy was plain. They were to engage us hotly on both wings, and then with a sufficient force carry the railroad, which, when gained, would cut our lines in two and be equivalent to flanking us right and left. No troops were at any time stationed along the line from the extreme left of the Twenty-sixth Regiment to the brick kilns, a distance of over 200 yards, until Colonel Avery ordered Major Lewis with four companies of the Thirty-third Regiment, to occupy it. There were also no troops defending the line from the brick kiln to where the main breastworks touched the railroad, a distance of 200 yards or more.

"The enemy now determined to carry this part of the line of our defence. What part the militia, who were stationed along the main breastworks nearest the railroad, and the Thirty-fifth Regiment, who were next to them, took in resisting this attempt, I cannot say. The brick kilns and other buildings excluded the view. These troops were certainly near enough, and by a proper change of front could have thrown themselves upon the enemy and overwhelmed him.

"The force attempting this point of our works, I do not believe to have been more than one regiment. (It was only the right wing of the Twenty-first Massachusetts Regiment), and the main resistance he encountered came from the rifle pits occupied by Major Carmichael's and Major Lewis' commands. The enemy was held in check for some considerable time, but at last he succeeded and carried the railroad between the brick kilns and the main breastworks and a part of his force passed in. They had advanced but a short distance, however, when they were met by a part of the Seventh North Carolina Regiment and driven out at the point of the bayonet, the Yankees leaping over the breastworks into the ditch beyond.

"It was during this time that we met with a severe loss in the death of Major Carmichael—as true a patriot and as brave a gentleman as ever lived. His death occurred in this manner: Colonel Avery and Major Carmichael were standing together at the corner of the traverse nearest the railroad. They were watching the action on the left and beyond the brick yard, when a single ball, whether aimed at the party or not, entered the mouth of Major Carmichael as he was speaking, and passed out at the back of the neck. I was standing at his side when he fell. He died instantly. A feeling of bitter grief ran through the trenches as he fell, for there was not a man in the Twenty-sixth regiment who was not devotedly attached to him. During the battle, Major Carmichael wore a small Confederate flag, perhaps three by four inches in dimension, mounted on a staff and attached to his cap. This may have attracted the fatal shot." The flag had been given the Major by a lady of New Bern, and he had promised her he would wear it in his cap in his first battle. It was doubtless the cause of his being singled out by some sharpshooter.

We will now return to the left of the Confederate line between the railroad and Fort Thompson. General Branch's paucity of troops prevented his taking advantage of Lieutenant-Colonel Haywood's brilliant bayonet charge with the Seventh Regiment. The enemy were driven back, but there were no soldiers to occupy the vacant line of defense at the brick yard, or to take the place in the works vacated by the retreat of the militia and the Thirty-fifth Regiment. Says General Branch, in his report: "The whole of the militia had abandoned their positions. Colonel Sinclair's Regiment very quickly followed their example. This laid open Haywood's right and a portion of the breastworks was left vacant. I had not a man with whom to occupy it, and the enemy soon passed in a column along the railroad and through a portion of the cut down timber in front which marched up behind the breastworks to attack what remained of Colonel Campbell's command." How this was done we will explain by quoting from Brigadier-General Parke, commanding the force supporting Reno's Brigade attacking the Confederate right wing.

"Lieutenant-Colonel Clark, commanding the Twenty-first Massachusetts, meeting Colonel Rodman, of the Fourth Rhode Island, informed him he had been in the works and assured him of the feasibility of again taking the intrenchments.

"I approved of this course on the part of Colonel Rodman, and at once ordered the Eighth Connecticut and the Fifth Rhode Island to his support. Passing quickly by the rifle pits which opened on us with little injury, we entered in rear of the intrenchments and the regiments in a gallant manner carried gun after gun, until the whole nine brass pieces on their front line were in our possession, the enemy sullenly retiring, firing only three guns from the front and three others from the fort on their left (Thompson). The Eighth Connecticut and Fifth Rhode Island followed immediately in the rear, and in support of the Fourth Rhode Island. Although now in possession of the entire works of the enemy between the railroad and the river, the heavy firing on our left and beyond the railroad proved that General Reno's Brigade was still hotly engaging the enemy.

"I ordered the Fifth Rhode Island Battalion and the Eighth Connecticut to advance cautiously. Captain J. N. King then reported that the enemy still occupied rifle pits along side the railroad back of the brick yard and a series of redoubts extending beyond the railroad and in General Reno's front.

"I then had the Fourth Rhode Island brought up and ordered the Colonel to drive the enemy from his position. This order was executed in a most gallant manner. The regiment charged the enemy in flank, while a simultaneous charge was made by General Reno in front, thus driving the enemy from his last stronghold."

General Burnside in his report of the battle, says: "General Foster seeing our forces inside the enemy's lines, immediately ordered his brigade to charge, when the whole line of breastworks between the railroad and the river were most gallantly carried. After the cheers of our men had subsided, it was discovered that General Reno was still engaged with the enemy on the left, upon which General Parke moved back with a view of getting in rear of the enemy's forces in the intrenchments to the left of the railroad. General Foster, also moved forward with one of his regiments, with a view of getting to the rear." It was to this last regiment that Colonel Avery and Captain Rand surrendered. This regiment General Foster marched down the county road leading to New Bern, until opposite the camp of the Twenty-sixth Regiment, when turning to the left, he marched through the woods and took position on both sides of the railroad; he also brought up four pieces of artillery and placed them in position.

Let us now return to Captain Rand's account of the closing incidents of the battle on his part of the line: "The action at this place had now continued for more than three hours. Our men from first to last poured in their fire with deliberate aim. Colonel Avery was everywhere along the trenches animating the men by his presence. I may say that nearly every man did his duty nobly. Many were the narrow escapes. Colonel Avery received a ball through his cap, and many received balls through their hats or clothes. The respective forces were all the time within from two to three hundred yards of each other; all had been silent along our lines, both right and left of us for some time. Just at this time, while we were so intently engaged on our front, we were fired into on our left by a considerable body of the enemy who had taken position in the edge of the woods beyond the railroad. This determined the conflict so far as we were concerned. Colonel Avery saw in an instant that nothing now remained but to draw off the troops. The order was given and we went out of the trenches amidst a perfect storm of bullets from both right and left.

The intention of Colonel Avery was to rally the men and form line on the railroad. He succeeded in a great measure, and marched diagonally through the woods, a distance of three or four hundred yards, for a point on the railroad just above the camp of the Twenty-sixth Regiment. My company occupied the extreme left of the rifle pit, and became the right of the line in retreat. The woods were so filled with underbrush that we could see but a short distance before us. When we had advanced far enough to see through the opening made for the railroad track, and had nearly reached the place where we were to form line, we discovered just across the railroad, and about fifty or seventy-five yards in front of our right, four pieces of the enemy's artillery and a regiment of infantry deployed on each side and extending across the railroad. An officer immediately rode out and demanded a surrender. Seeing ourselves surrounded and no hope of escape, Colonel Avery, and those on the right, surrendered. Those on the left, being further off, and aided by the cover of the woods, nearly all escaped. The surrender took place at 11:30 o'clock a. m. The number of prisoners taken at this place were about one hundred and fifty. The number of prisoners taken in all were two hundred and six." This admirable and intelligent account of the battle was prepared by Captain Rand, shortly after his capture. It is accompanied with a diagram of the battle field made by Lieutenant Woodbury Wheeler, of Latham's Battery, who was also captured.

These gentlemen were permitted to visit the battle field from one end to the other, and they carefully made notes for the purpose of giving an account of the battle. Space forbids my quoting the report in its entirety. I will only make one further quotation: "We received no orders to retreat, neither did we receive orders of any kind during the whole course of the battle. The woods were very thick, which, coupled with the mist of the morning, made it impossible to see our troops on either side. We retreated because we were exposed to a cross fire, and because it would have been certain destruction to have held our places five minutes longer. No officer or man dreamed of such a thing as being taken prisoner. We could have made good our retreat if we had received the order as others did."

In justice to General Branch, on this point, I quote from his official report: "Finding the day was lost, my next care was to secure the retreat. I dispatched two couriers to Colonel Avery and two to Colonel Vance, with orders to fall back to the bridges, etc., etc." These couriers never delivered their orders. This account will be incomplete without making quotation from Colonel Vance's and Lieutenant-Colonel Robert F. Hoke's reports of this battle. Lieutenant-Colonel Hoke says: "The regiment moved up to the scene of action in fine style, Colonel Avery in command in the centre, I of the right wing, Major Lewis of the left. Colonel Avery gave the command to fire, which seemed to have great effect, as the enemy scampered. Major Lewis then moved to the right of the railroad with several (four) companies, and engaged the enemy from that time until after 12 o'clock. He behaved most gallantly, was in the hottest of the whole battle field. He repulsed the enemy time and again, and twice charged them with detachments from his companies, and each time made them flee. Our loss was greater at that point than any other, as he had to fight to his front, right, and left, but still maintained his position. Finding the enemy were getting in strong force on our right, and were going to turn our right flank, as there were no troops between our regiment and the left of Colonel Vance's companies, a distance of a quarter of a mile, I moved quickly with Captain Park's company, and sent a messenger to Colonel Avery for another company. He immediately sent me Captain Kesler's company. I ordered the whole to fire, which did great execution, as the enemy fell and fled, but soon appeared in strong force and again we drove them back, but soon they again appeared in stronger force, and engaged us, which continued until 12:30 o'clock. At 12:15 o'clock I saw a United States flag flying upon one of our works, but saw Colonel Avery still fighting. I did not know that Colonel Avery and Major Lewis had fallen back until I saw the enemy upon my left with several regiments, and about fifty yards to the rear of the position Colonel Avery had occupied. I ordered the men under my command to fall back, but to do so in order. We were hotly fired at as we fell back."

I next quote from Colonel Z. B. Vance's report of the battle: "The regiment was posted by Lieutenant-Colonel Burgwyn in the series of redans, constructed by me on the right of the railroad, in the rear of Bullen's Branch, extending from the railroad to the swamp, about 500 yards from the road by Weathersby. At this road I had constructed the night before a breastwork, commanding the passage of the swamp, and there was placed a section of Brem's artillery, Lieutenant Williams commanding, and Captain McRae's company of infantry, with a portion of Captain Hays' and Lieutenant W. A. Graham's Second Cavalry (Nineteenth North Carolina) dismounted. About 2 o'clock Friday morning (14 March) I pushed Companies B, E, and K, of my right wing across the small swamp alluded to so as to make my extreme right rest on the battery at the Weathersby road. During the day, two companies of the Thirty-third Regiment, under Lieutenant-Colonel Hoke, about 9 a. m., were placed in the redans vacated by my right companies.

The battle began on my left wing about 7:30 a. m., extending towards my right by degrees, until about 8:30 a. m., all the troops in my command were engaged as far as the swamp referred to.

The fight was kept up until about 12 o'clock, when information was brought me by Captain J. J. Young, my Quartermaster, who barely escaped with his life in getting to me, that the enemy in great force had turned my left by the railroad track at the woods and the brick yard, had pillaged my camp, were firing in reverse on my left wing, and were several hundred yards up the railroad between me and New Bern. Also that all the troops were in full retreat except my own.

Without hesitation, I gave the order to retreat. My men jumped out of the trenches, rallied and formed in the woods without panic or confusion, and having first sent a messenger to Lieutenant-Colonel Burgwyn to follow with the forces on the right, we struck across the Weathersby's road to Bryce's Creek. On arriving at the creek, found only one small boat, capable of carrying only three men. The creek here is too deep to ford and seventy-five yards wide. Some plunged in and swam over, and swimming over myself, I rode down to Captain Whitford's house on the Trent river, and through the kindness of Mr. Kit. Foy, procured three more small boats. Carrying one on our shoulders, we hurried up to the crossing. In the meantime, Lieutenant-Colonel Burgwyn arrived with the forces of the right wing in excellent condition, and assisted me with the greatest coolness and efficiency in getting the troops across, which, after four hours of hard labor, and the greatest anxiety, we succeeded in doing. Lieutenant-Colonel Burgwyn saw the last man over before he entered the boat. I regret to say that three men were drowned in crossing.

"A large Yankee force were drawn up in view of our scouts, about one mile away, and their skirmishers appeared just as the rear got over."

Of the deaths of Major Carmichael and Captain Martin, Colonel Vance thus feelingly speaks:

"Major A. B. Carmichael fell about 11 a. m. by a shot through the head, while gallantly holding his post on the left, under a most galling fire. A braver, nobler soldier never fell on field of battle. Generous and open-hearted, as he was brave and chivalrous, he was endeared to the whole regiment. Honored be his memory. Soon thereafter, Captain W. P. Martin, of Company H, also fell, near the regimental colors. Highly respected as a man, brave and determined as a soldier, he was equally regretted by his command, and by all who knew him. Lieutenant Porter, of Company A, was also left behind wounded. Captain A. N. McMillan was badly wounded, but got away safely.

"Once across Bryce's Creek, we were joined by Lieutenant-Colonel Hoke, Thirty-third Regiment, with a portion of his command, and took the road for Trenton. We marched night and day, stopping at no time for rest or sleep more than four hours. We arrived at Kinston safely about noon on 16 March, having marched fifty miles in about thirty-six hours."

"I cannot conclude this report," says Colonel Vance, "without mentioning in terms of the highest praise the spirit of determination and power of endurance manifested by the troops during the hardships and sufferings of our march. Drenched with rain, with blistered feet, without sleep, many sick and wounded, and almost naked, they toiled on through day and all the weary watches of the night without murmuring, cheerfully, and with subordination, evincing most thoroughly the high qualities in adversity which military men learn to value even more than courage on the battle field."

We close this account of the battle with one or two incidents. When Bryce's Creek was reached, there was some confusion, and a natural eagerness to get across, as the enemy's guns were heard in the distance. Many attempted to swim across, and several were drowned before the officers could restrain them. Colonel Vance, to inspire confidence, spurred his horse in the creek, the animal refusing to swim, the Colonel became unseated and weighed down with his accoutrements, he sank from view in the dark water of the stream and was about to be drowned, when assistance was rendered him, and he reached the opposite side in safety. Lieutenant-Colonel Burgwyn and his college-mate, Lieutenant W.

A. Graham, Company K, Nineteenth North Carolina (Second Cavalry), taking their stand on opposite sides of a path leading to the stream, with swords crossed, counted the men off in boat load lots as they were called out, and in this way without confusion or crowding, all were successfully ferried over and these two officers were the last to step aboard.

Major Wm. A. Graham, so widely known in the State for his prominence in agricultural matters, at the battle of New Bern was Lieutenant in command of Company K, Second North Carolina Cavalry, and the writer has been so fortunate as to get him for an eye witness account of that part of the battlefield where his command was posted, as follows:

"My company (K) was dismounted and placed in the brick yard. About sun set was ordered to report to Colonel Vance, Twenty-sixth North Carolina Troops, who sent me to Lieutenant-Colonel Burgwyn, commanding right wing of the Twenty-sixth Regiment and the companies on the road (Weathersby). Colonel Burgwyn placed my company on picket some half mile or more beyond the bridge, and he, with writer, scouted on flank of the pickets. The axes of the enemy could be heard cutting a road along the railroad.

"Next morning Captain Hayes, of Company A, Second Cavalry, reported. The pickets were called in and everything made ready for the battle. The forces at the road (Weathersby) consisted of Companies A and K, Second Cavalry, a section of the Charlotte battery, Lieutenant A. B. Williams in command and Captain McRae's independent company of infantry. Company K connected the force in the road with the right of the Twenty-sixth Regiment. No enemy appeared in our front and when Colonel Burgwyn began forming the companies of the Twenty-sixth in rear of the entrenchments, we had no idea we had been defeated, but thought it was probably for pursuit. Going to him for orders, he informed me that we had been defeated on the left and must try and beat the enemy to New Bern.

"Everything moved off in fair order until getting near the crossing of the railroad, a scout announced the enemy coming up the railroad only a short distance off. Colonel Burgwyn ordered the artillery and Captain Hayes' company, who were mounted, to save themselves, which they proceeded to do. Colonel Burgwyn, with the infantry, took to the left through the woods. He dismounted his orderly and gave me one of his horses and ordered me to scout to the left and forward to see if the bridges were standing. Coming out at the camp of the Thirty-seventh Regiment, I saw both bridges on fire and so reported. We then struck the trail of Colonel Vance's retreat and overtook his command at Bryce's creek, endeavoring to cross in a boat, carrying three men. Colonel Vance had swam his horse across the creek and had gone to hunt other boats. It was reported that the enemy were close upon us and at least half of the men threw their arms in the creek, saying they did not intend that the Yankees should have them. There was great confusion. Colonel Burgwyn was as cool as if nothing unusual was transpiring. Calling such of the officers as he saw to him, he announced he would hold a "council of war," told the council we were responsible for the action of the men, and must form them and keep order. This was done. Men were sent up and down the creek to hunt boats.

"In the afternoon a negro man who belonged to —— Foy, came to the opposite side of the creek and announced there was a boat a mile or so down the creek where Colonel Hoke (R. F.) had crossed. The men moved off through the swamp down the creek, sometimes up to the armpit in the mire. The negro went along on the other side, and when he reached the boat he halloed and we went to him. I got into the boat and had just taken a seat, when Colonel Burgwyn called me to him and said I must help him keep the men from overloading and sinking the boat; the boat would hold eighteen. I stood facing Colonel Burgwyn, and each time as we counted eighteen we halted the column. When we all had crossed except Colonel Burgwyn and myself, I entered the boat and, leading the horse into the water, swam him over along its side. The boat returned and Colonel Burgwyn came over in like style. It was now near sun set. Colonel Burgwyn took command of such formation as there was and held it until we reached Trenton next day, where we found Colonel Vance and several hundred men of the different commands which had been at New Bern. Colonel Vance assumed command and brought the troops to Kinston."

When Captain J. J. Young met the fleeing militia, he tried to rally them—exhorted them to go back and rejoin their comrades fighting in the works, saying, their conduct would forever disgrace them; that the papers would be full of their cowardice, etc., etc. One of them replied: "I had rather fill twenty newspapers than one grave." Some of the militia did not stop running until they reached New Bern. One was found dead on the rear platform of the last train as it crossed the river into New Bern, expiring as he reached the train just starting, having run all the way from the battle field, about five miles.

To make this account historically complete, I append list of the troops engaged on either side, and the casualties sustained.

CONFEDERATE FORCES, BRIGADIER GENERAL L. O'B. BRANCH, COMMANDING.

	REGIMENTS.									
	7 N. C.	19 N. C. (Cavalry.)	26 N. C.	33 N. C.	27 N. C.	35 N. C.	37 N. C.	Brem's	Latham's.	Totals.
Killed	6	0	5	32	4	5	1	1	10	64
Wounded	15	0	10	28	8	11	3	3	11	89
Missing and Prisoners	30	0	72	144	42	9	8	8	22	335
										488

UNION FORCES, BRIGADIER GENERAL A. E. BURNSIDE, COMMANDING.

	BRIGADES.			
	Foster's Brigade, 23, 24. 25 and 27 Mass. and 10 Conn.	Reno's Brigade, 21 Mass., 51 N. Y., 9 N. J. and 51 Pa.	Parke's Brigade, 4 R I., 5 R. I., 8 and 11 Conn.	Totals.
Killed	37	30	21	88
Wounded	145	167	58	370
Artillery	2 killed, 8	wounded.		10
				465

So much space is given to the account of this, the *first battle* in which the regiment was engaged, because it was its first battle, and the conduct of its officers and men was so altogether creditable. No troops could have borne themselves better under the ordeal to which they were exposed. The rapidity of General Burnside's advance took General Branch by surprise. The latter expected at least a day's delay at Fisher's landing, and at the Croatan breastworks above Otter Creek, but there was no fight at these advanced points of defense, and nothing delayed the enemy's rapid approach. Another day and the brick yard would have been defended by artillery, and this point secure, General Burnside would have failed in his attempt to capture New Bern. The disparity of forces was great, but General Foster, with his five regiments, opposed by Colonels Campbell and Lee, with their three, could make no headway on the Confederate left; and General Reno, with his four regiments, assisted by General Parke, was regularly driven back by the Twenty-sixth and Thirty-third Regiments on the right. One regiment to have replaced the 350 militia, and the Thirty-fifth Regiment, would have stood as firm as the others, and there would have been no undefended part of the line to let the enemy through; and reinforcements, which were hurrying to General Branch's assistance, would have reached him during the day.

General Burnside well won his promotion as Major-General, which was the result of his victory, whereas on the Confederate side, this battle introduced to the military world names to become distinguished in the annals of the war.

The press of the State heaped eulogies upon the officers and men of the Twenty-sixth Regiment and recruits flocked to its standard.

Governor Vance applied for and received permission to recruit his regiment to a legion, and was in a fair way to succeed, several companies having arrived in camp, and others were at home drilling, when he gave up the attempt in disgust at what he thought was "the opposition to the scheme on the part of the State and Confederate authorities," and the companies were disbanded.

While resting at Kinston, after the battle of New Bern,

Captain N. P. Rankin, of Company F, was elected Major vice Carmichael, killed; and First Lieutenant Clement Dowd elected Captain of Company H, vice Martin, killed; First Lieutenant Joseph R. Ballew was promoted to be Captain of Company F.

The troops around Kinston were now reorganized. Brigadier-General French, on 16 March, reached Goldsboro and relieved General Branch of the command of the District of Pamlico; and 19 March General Gatlin was relieved of command on account of ill health, and Major-General Theo. H. Holmes, assigned to the command of the Department of North Carolina. On 17 March Brigadier-General Robert Ransom was ordered to Goldsboro "for duty with troops in the field," and a brigade was formed for him consisting of the Twenty-fourth, Twenty-fifth, Twenty-sixth, Thirty-fifth, Forty-eighth and Forty-ninth North Carolina Regiments. Under this gallant and accomplished soldier and disciplinarian, numerous drills and strict camp regulations prevailed until on 20 June, 1862, the brigade was ordered to Virginia to join Lee's army, then confronting McClellan below Richmond.

REORGANIZATION FOR THE WAR.

The Twenty-sixth Regiment was a twelve-months regiment, and in the Spring of 1862 re-enlisted for the war. The men in the ranks were given the right to elect their company officers, and the latter the right to elect field officers.

Many changes took place in the regiment at its reorganization. Colonel Vance was always most popular with his men. He sought and obtained to the fullest extent the love of his soldiers, was always solicitous of their welfare and comfort, leaving chiefly to his second in command matters of drill and discipline. At no time was there any doubt as to his re-election.

As to Lieutenant-Colonel Burgwyn, had the election taken place before the regiment had in actual battle experienced the benefit of drill and strict obdience to orders, he could not have been re-elected. Says an officer of the regiment (Captain Thomas J. Cureton): "Colonel Burgwyn was emphatically a worker in camp, careful of the comforts of his men, constantly drilling; he believed in discipline and endeavored to bring his regiment to the highest state of efficiency. I always found him strict in camp, so much so, that up to the battle of New Bern he was very unpopular, and I often heard the men say if they ever got into a fight with him what they would do, etc., etc."

The morning before the fight, Burnside's gunboats were coming up the river, shelling the banks. Colonel Vance was placed in command of the right of our line, or in other words, acting Brigadier-General. Lieutenant-Colonel Burgwyn was, therefore, in command of the Twenty-sixth Regiment. He suspected the feelings of the men towards him. He formed the regiment at the point where the breastworks crossed the railroad, and addressed them in substance as follows: "Soldiers! the enemy are before you, and you will soon be in combat. You have the reputation of being one of the best drilled regiments in the service. Now I wish you to prove yourselves one of the best fighting. Men, stand by me, and I will by you." The response was unanimous—"We will," from the men. Next day the battle was fought. Only the left companies of the regiment under the command of Major Carmichael, and Captains Rand and Martin were most actively engaged, and suffered heavily. The right companies, when they found the enemy on their flank and getting in their rear, had to fall back to find the bridge across the Trent, on fire, our troops all gone, and the only way of escape was to cross Bryce's Creek.

When we got there only a small boat that would carry two people at a time could be found. Colonel Vance rode his horse in the creek, which refused to swim, and the colonel was very nearly drowned before assistance reached him. Several of the men were drowned trying to swim the creek. When the boat reached the bank we were on, an officer called to Colonel Burgwyn to get in first. He was met with the reply: "I will never cross until the last man of my regiment is over." Nor did he till the *last man* was over.

We retreated up to Trenton Court House and expected pursuit. Colonel Burgwyn was always in the rear. From this time on he had the entire confidence of his men and was their pride and love. Colonel Vance and Lieutenant-Colonel Burgwyn received practically the unanimous vote of the regiment.

CHANGES IN THE OFFICERS AT REORGANIZATION.

First Lieutenant James S. Kendall, Company K, was elected Major. This gallant officer and accomplished soldier only survived his promotion a few weeks, dying before the regiment left for Virginia, from yellow fever, contracted at Wilmington while on furlough.

First Lieutenant William Wilson became Captain of Company B; Second Lieutenant James T. Adams, Captain of Company D; Second Lieutenant John T. Jones, Captain of Company I; Second Lieutenant John C. McLauchlin, Captain of Company K., and First Lieutenant S. W. Brewer, Captain of Company E.

A WOMAN RECRUIT.

While the Twenty-sixth Regiment was in camp in and around Kinston, after the battle of New Bern, many recruits joined the command. Among them were two young men, giving their names as L. M. and Samuel Blalock. They enlisted in Captain Ballew's company (F) and were brought to the regiment by private James D. Moore, of Company F. On the way from their home, in Caldwell County, to join the regiment, Moore was informed in strict confidence by L. M. (Keith) Blalock, that Samuel was his young wife, and that he would only enlist on condition that his wife be allowed to enlist with him. This was agreed to by Moore, who was acting as recruiting officer, and Moore also promised not to divulge the secret. Sam Blalock is described as a good looking boy, aged 16, weight about 130 pounds, height 5 feet and 4 inches, dark hair; her husband (Keith) was over 6 feet in height. Sam Blalock's disguise was never penetrated. She drilled and did the duties of a soldier as any other member of the company, and was very adept at learning the manual and drill.

In about two months her husband, who was suffering from

hernia and from poison from sumac, was discharged, and Sam informed his Captain and Colonel Vance, that he was a woman, whereupon she was discharged and permitted to join her husband.

On returning home, Keith Blalock and his wife, now known by her real name, "Malinda," joined Kirk's command, an organized body of Union troops, made up largely of deserters and bushwhackers, operating in the Western part of the State.

In the Spring of 1864, while the said James D. Moore was at home at his father's, at a place called the Globe, recovering from the wound he had received at Gettysburg, the house was attacked by Keith and Malinda Blalock, and their gang, and Carroll Moore, his father, severely wounded. Several of the marauders were wounded, and among them Malinda.

Again in the fall of 1864, Keith and his raiders attacked Mr. Carroll Moore's house, and were again driven off. This time Keith was shot in the head, and one eye put out.

After the war, Keith attempted merchandizing in Mitchell County and was a candidate for the Legislature on the Republican ticket, but was defeated, and about 1892 he and his wife went to Texas. They subsequently returned to North Carolina, and at this time (1901) are living in Mitchell county. Malinda Blalock's maiden name was Pritchard, and her brother, Riley Pritchard, was United States Commissioner in President Harrison's Administration.

MALVERN HILL, JULY 1, 1862.

Ordered to Virginia, 20 June, 1862. Ransom's Brigade was directed to report to General Huger on the Williamsburg road, and a little before dark on the night of 25 June, Colonel Vance's Regiment relieved the Twenty-fourth North Carolina Regiment on picket duty in front of the enemy.

The night was very dark, and with no one to direct them, the regiment took position on one side of a rail fence and in front of a hedge row. As it happened, the enemy were lying down in line of battle on the opposite side, and abiding their time. After the Twenty-sixth had gotten quieted down for the night, in entire ignorance of the presence of the enemy, the latter suddenly arose, thrust their guns through the fence rails and opened fire. So close were they to us, says a member of the regiment, that the beards of many of the men were singed. The surprise was so great that seven of the companies on the right of the regiment went to the rear; however, Companies G, H and K, undaunted by the nearness and numbers of the enemy, remained on the field. On the next morning those companies were highly complimented by their field officers for their exceedingly creditable conduct in holding their lines during the night under such trying circumstances. Again, on picket, on the 27 June, the Twenty-sixth Regiment was pushed to the front and took possession of some unfinished works of the enemy. Just as it was about to be relieved, it was attacked, but returned the fire so briskly and with such effect as to drive the enemy back.

Quoting from so much of Brigadier-General Robert Ransom's report of the part his brigade took in the battle of Malvern Hill, as applies to the Twenty-sixth Regiment, he says: "At 7 p. m. (July 1, 1862) I received the third request from General Magruder, that he must have aid, if only one regiment. The message was so pressing that I at once directed Colonel Clarke to go with his regiment (Twenty-fourth North Carolina). The brigade was at once put in motion, Colonel Clarke had already gone, Colonel Rutledge next, then Colonel Ransom, Colonels Ramseur and Vance, all moved to the scene of conflict at the double quick. As each of the three first named regiments reached the field, they were at once thrown into action by General Magruder's orders. As the last two arrived, they were halted by me to regain their breath, and then pushed forward under as fearful fire as the mind can conceive.

"Ordering the whole to the right so as to be able to form under cover, I brought the brigade in line within 200 yards of the enemy's batteries. It was now twilight; the line was put in motion and moved steadily forward to within less than 100 yards of the batteries. The enemy seemed unaware of our movements. Masses of his troops appeared to be moving from his left towards his right. Just at this instant the brigade raised a tremendous shout, and the enemy at once wheeled into line and opened upon us a perfect sheet of fire from muskets and the batteries. We steadily advanced to within twenty yards of the guns. The enemy had concentrated his forces to meet us. Our onward movement was checked; the line wavered and fell back before a fire, the intensity of which is beyond description. It was a bitter disappointment to be compelled to yield when their guns seemed almost in our hands."

The losses sustained by Ransom's Brigade from 26 June to 1 July, 1862, inclusive, embraced three Colonels wounded, one Lieutenant-Colonel killed, several field officers and many company officers, and a total of 499 privates killed and wounded.

Casualties separately stated:

Regiments	24th.	25th.	26th.	35th.	49th.
Killed	9	22	6	18	14
Wounded	42	106	40	91	75

INCIDENTS OF THE BATTLE.

During the charge of the regiment at Malvern Hill, Captain Lane, of Company G, had the pocket of his coat cut open by a ball, and the contents fell on the ground. Among these was a package wrapped in newspaper, containing the month's pay of his company. Next morning Captain Lane discovered his loss, obtained permission to go and hunt for it, and strange to say, found the package untouched, lying in the open ground where it had fallen among the dead and wounded.

After the regiment had taken its position for the night after the charge, and the officers and men were resting on their arms, Captain Lane lay down between two of his soldiers and fell asleep. Next morning when he awoke the man on his right and left had both been killed by the enemy's fire while asleep, and their deaths not discovered. They awoke to the sound of the "reveille" in another world.

While the men were lying down in line of battle, waiting the order to charge, they were subjected to a furious shelling, and there was more or less dodging of the head as the missiles whizzed by. "Why are you so polite in the presence of the enemy," remarked Colonel Vance. A rabbit was flushed by the line as it advanced, which caused the men to raise a shout as it ran past them, whereupon Colonel Vance joined in the cry, saying: "Go it cotton tail. If I had no more reputation to lose than you have, I would run too."

On 7 July Ransom's Brigade was ordered back to General Holmes' command, and on 31 July, 1862, Major-General D. H. Hill relieved General Holmes in command of the Department of North Carolina, and 11 August Brigadier-General J. Johnston Pettigrew, who had been severely wounded and captured at the battle of Seven Pines, 1 June, 1862, was assigned to the command of Petersburg, and given the brigade then under the command of Colonel Junius Daniel.

TWENTY SIXTH REGIMENT DETACHED FROM RANSOM'S AND ASSIGNED TO PETTIGREW'S BRIGADE.

Colonel Vance's election as Governor in August, 1862, caused a vacancy in the Colonelcy of the Twenty-sixth Regiment. The Lieutenant-Colonel was not 21 years of age, and the opposition of General Ransom to his promotion on account of his age, the General saying: "He wanted no boy Colonel in his brigade," was well known to the regiment, and indignantly resented.

Application was made through the proper channels for a transfer to some other brigade, and on 26 August, 1862, by special order No. 199, from the A. & I. G. office, at Richmond, the Twenty-sixth Regimont was detached and ordered to report to Brigadier-General S. G. French, at Petersburg, Va., for duty with the brigade formerly commanded by Brigadier-General J. G. Martin.

Referring to the election of Colonel Vance as Governor, one of the regiment writes as follows: "Though rejoicing that he had been chosen Governor of the State by such a complimentary majority, with a pang of regret we saw Colonel, now Governor-elect Z. B. Vance, exchange the sword for the helm of State. He received almost the unanimous support of the regiment, there being only seven votes cast against him, which well attests his popularity among his troops.

"His separation from us was quite sad, all feeling the heavy loss to the regiment. In his farewell address to the regiment, he, with his usual truthfulness and sincerity, scorned to hold out any false promises to those who had been under his command, telling them plainly, that all they could expect was 'War! War!! War!!! Fight till the end.'

"But in the promotion of Lieutenant-Colonel Burgwyn to the Colonelcy of the regiment, we gained an officer, young, gallant and brave, and eminently fitted to fill the vacancy."

Speaking of the transfer of the regiment to Pettigrew's Brigade, this writer goes on to say: "Never was there a more fortunate change. It seemed as if Pettigrew and Burgwyn were made for each other. Alike in bravery, alike in action, alike in their military bearing, alike in readiness for battle and in skillful horsemanship, they were beloved alike by the soldiers of the Twenty-sixth. Each served as a pattern for the other, and in imitating each other they reached the highest excellence possible of attainment in every trait which distinguishes the ideal soldier." It will be of pathetic interest to state in addition to the above eloquent panegyric, that both General Pettigrew and Colonel Burgwyn were alumni of the State University, and fell on the field of battle within a few days of each other, the one on Gettysburg's gory field, 1 July, 1863; the other, commanding the rear guard of the army on its retreat across the Potomac at Falling Waters, 14 July, 1863.

The promotion of Lieutenant-Colonel Burgwyn, and the death of Major Kendall, who had been sick since his election, required the filling of the positions of Lieutenant-Colonel and Major. A board of examination having been appointed to pass upon the qualifications of all officers before their promotion, Captain John R. Lane, of Company G, and Captain John T. Jones, of Company I, were summoned for examination, and obtaining the favorable report of the board, which was composed of Colonel H. K. Burgwyn, of the Twenty-sixth; Colonel Thomas Singletary, of the Forty-fourth, and Lieutenant-Colonel T. L. Hargrave, of the Forty-seventh North Carolina Regiments, duly received their commissions as Lieutenant-Colonel and Major, respectively. About this time, Captain Ballew, of Company F, resigned and First Lieutenant R. M. Tuttle was promoted to be Captain of this company, to become famous above all other companies in the army, from the fact that every member present, numbering ninety-one, was killed or wounded in the battle of Gettysburg. Captain Steele, of Company B, also resigned, and First Lieutenant Thomas J. Cureton became Captain, and served most gallantly to the end. Lieutenants H. C. Albright and N. G. Bradford were promoted to be Captains of Companies H and I, respectively.

PETTIGREW'S BRIGADE.

This brigade to become so famous in military annals, was composed of the Eleventh, Twenty-sixth, Forty-fourth, Forty-seventh and Fifty-second North Carolina Regiments.

Of the commander of this brigade, later on in this sketch a more extended notice will be given. He was, at the time of its organization, convalescent from the severe wound received on 1 June, 1862, at the battle of Seven Pines, and was placed in command of Petersburg in the fall of 1862. During the months of September, October, November and December, 1862, Pettigrew's Brigade was either on duty in Virginia or North Carolina.

The faithfulness with which Colonel Burgwyn disciplined the regiment, much improved its efficiency, and it became known as one of the best drilled regiments in the service. In his labors in this behalf, he was ably seconded by his Lieutenant-Colonel, John R. Lane, who manifested extraordinary abilities as a drill master, and disciplinarian. "This perfection of drill, to which the excellent music of Captain Mickey's band greatly added, was a cause of just pride to every member of the regiment, officers and men alike. Never was any man prouder of his regiment and of his band, considered the finest in the army of Northern Virginia, than Colonel Burgwyn," writes a member of the regiment.

RAWLS' MILLS, 2 NOVEMBER, 1862.

The first opportunity afforded the Twenty-sixth regiment to show of what stuff it was made, acting in an independent command, occurred in the engagement at Rawls' Mills, in Martin County, N. C., in resisting General J. G. Foster's attempt to capture the regiment while on a reconnoissance in the neighborhood of Washington, Beaufort County.

In his report of the expedition, General John G. Foster, commanding the Federal troops in North Carolina, with headquarters at New Bern, says he set out on 31 October, 1862, from New Bern to capture the three regiments (Seventeenth, Twenty-sixth and Fifty-ninth North Carolina) foraging through the Eastern counties of the State. He took three brigades, 21 pieces of artillery and cavalry, with ample wagon train, total 5,000 men.

On 2 November, 1862, Foster left Washington for Williamston. On the same evening he encountered the Twenty-sixth Regiment at Little Creek. He says: "I ordered Colonel Stevens, commanding Second Brigade, to drive them away. The engagement lasted one hour, when the enemy being driven from their rifle pits by the effective fire of Belger's Rhode Island Battery, retired to Rawls' Mill. One mile further on, where they made another stand in a recently constructed field work, Belger's battery and two batteries of the Third New York artillery, after half an hour, succeeded in driving the enemy from their works, and across the bridge, which they burned. We bivouacked on the field, and next day proceeded to Williamston."

The only Confederate troops to oppose these 5,000 of Foster were six companies of the Twenty-sixth Regiment, under Colonel Burgwyn. Leaving four companies under Lieutenant-Colonel Lane, at Williamston, on the Roanoke river, Colonel Burgwyn started out on a reconnoissance to go as far as Washington, N. C. He stationed two companies at Rawls' Mills, under Captain McLauchlin, of Company K, with orders to fortify the position and proceeding with the remaining four, reached the vicinity of Washington, N. C., just as General Foster was starting out to capture him.

Colonel Burgwyn had no cavalry or artillery. There were two parallel roads leading out of Washington for Williamston. Again, it was necessary to delay the Federal advance as much as possible, to give time to Colonel Ferebee, of the Fifty-ninth Regiment (Fourth Cavalry) and Lieutenant-Colonel Lamb, in command of the Seventeenth Regiment, who were in the neighborhood of Plymouth, to retrace their steps. Dispatching a messenger to Colonels Lamb and Ferebee, warning them of their danger, and one to Lieutenant-Colonel Lane, with an order to join him at Rawls' Mills, Colonel Burgwyn determined to resist Foster's advance at that point.

As soon as it was ascertained which of the two roads the enemy had selected, Colonel Burgwyn chose the other and started out in the race for Rawls' Mills. On reaching the Mills, he ordered Captain McLauchlin to go down the road on which Foster was advancing, and hold him in check at Little Creek. Captain McLauchlin, with Companies K and I, reached Little Creek just as the enemy's cavalry began to cross, and attacked them with his handful of men.

Colonel Burgwyn, placing his four companies in the hastily constructed breastworks at the Mills, awaited Foster's advace. After Captain McLauchlin had been for some time engaged with the enemy at Little river, successfully defending the passage of the stream against Colonel Stevenson's brigade with cavalry and artillery, Colonel Burgwyn sent Companies D and F, under command of Major Jones, to Captain McLauchlin's support. Fearing that a longer resistance by so small a force would result in its capture, Colonel Burgwyn, after the fight had lasted over an hour, ordered Captain McLauchlin to join him at the Mills. Here General Foster brought into action three batteries of artillery against the six companies at the Mills, and succeeded, "according to the General's report," after half an hour, in driving the enemy from his works, and across the bridge, which they burned. The fact was, Colonel Burgwyn, having received advices that Colonels Ferebee and Lamb were safe, and Lieutenant-Colonel Lane having joined him from Williamston, concluded to retire in the night, so as not to disclose the paucity of his force, and at his leisure fell back in the direction of Tarboro, first burning the bridge at the Mill. Captain McLauchlin lost one killed, and three wounded.

General Foster's report admits a loss of six killed and eight wounded.

After proceeding to within ten miles of Tarboro, "owing to the exposed condition of his men and want of provisions," says General Foster, he abandoned any further advance, and countermarched to Washington, and thence to New Bern.

It was a singular coincidence that the Federal General (Foster) had been the tutor of his youthful antagonist (Burgwyn), when the latter was a student at West Point, in 1856, awaiting appointment in that institution, at which General Foster, then Captain Foster, was one of the professors. The art of war as taught by the professor was in this instance applied to his discomfiture by the pupil.

FOSTER'S EXPEDITION AGAINST GOLDSBORO.

In December, 1862, General Foster started out from New Bern to destroy the railroad bridge over the Neuse river, and capture Goldsboro, N. C. Major-General S. G. French, who was in command of the Department of North Carolina, under Major-General G. W. Smith, commanding at Richmond, assembled his forces to oppose him. On 17 December, 1862, a spirited engagement took place near Goldsboro, in which General Foster was driven back, and he hastily retreated to New Bern. Pettigrew's brigade was not seriously engaged in this action, but pursued General Foster on the latter's retreat.

GENERAL D. H. HILL'S ATTEMPT TO CAPTURE NEW BERN.

On 7 February, 1863, Major-General G. W. Smith resigned and Major-General D. H. Hill was again placed in command of the troops in North Carolina. General Hill resolved on the capture of New Bern. General Pettigrew was given command of the troops on the north side of the Neuse, and General Hill had charge of those to operate on the south side.

General Pettigrew with his brigade, started from Goldsboro on 9 March, 1863. By rapid marches he reached the enemy's works at Barrington's Ferry, near New Bern. The Twenty-sixth Regiment was ordered at daylight into position to carry the place. Three 20-pound Parrott guns relied upon to destroy the gunboats guarding the water approaches to New Bern, proved utterly worthless. One burst, the ammunition was defective and their fire proved more injurious to the Confederates than to the enemy. There was nothing to do but to withdraw. "The only question," says General Pettigrew in his report, "was whether I should carry the works before withdrawing. The Twenty-sixth Regiment had been in waiting ever since daylight, and would have done it in five minutes. The works we could not hold. There would be a probable loss of a certain number of men sixty miles from a hospital. I decided against it. It cost me a struggle after so much labor and endurance to give up the eclat, but I felt that my duty to my country required me to save my men for some operation in which sacrifice would be followed by consequences. I therefore withdrew the whole command except the Twenty-sixth Regiment, which remained within about 500 yards of the place, in order to cover the withdrawal of Captain Whitford's men. I cannot refrain from bearing testimony to the unsurpassed military good conduct of those under me. In seven days they marched 127 miles; waded swamps, worked in them by night and day, bivouaced in the rain, some times without fire, never enjoyed a full night's rest after the first, besides undergoing a furious shelling, and discharging other duties. All this without murmuring or even getting sick."

It was not long before General Pettigrew had another chance at the enemy, in which he was more fortunate. General Hill, with all his available forces, on 30 March, 1863, invested General Foster in Washington, N. C. On 9 April, 1863, at Blount's Creek, Pettigrew's brigade met and defeated General Spinola in the latter's attempt to raise the siege. Finding it impossible to capture the place after the enemy's gun boats had succeeded in passing the batteries at Rodman's Point, and thus reinforcing General Foster, after fourteen days investment, General Hill withdrew, having failed in this attempt to capture the town.

TWENTY-SIXTH REGIMENT.

1. James T. Adams, Lieut -Colonel.
2. Samuel P. Wagg, Captain, Co. A.
3. William Wilson, Captain, Co. B.
4. Stephen W. Brewer, Captain, Co. E.
5. Jos. R. Ballew, Captain, Co. F.
6. R. M. Tuttle, Captain, Co. F.
7. H. C. Albright, Captain, Co. G.

MAJOR GENERAL HARRY HETH'S DIVISION.

On 1 May, 1863, Pettigrew's Brigade was ordered to Richmond to be ever thereafter attached to the Army of Nothern Virginia. Taking position first at Hanover Junction, to protect that important point in the enemy's attempts to capture Richmond, the brigade, leaving the Forty-fourth Regiment behind at the junction, as a guard, proceeded to Fredericksburg, and now attached to Heth's Division, set out on 15 June on the memorable march to invade Pennsylvania.

Heth's Division, as then organized, was composed of Archer's Tennessee, Davis' Mississippi, Brockenborough's Virginia, and Pettigrew's North Carolina Brigades.

The division commander was a native of Virginia, a graduate of West Point, had served with distinction in the war with Mexico, and against the Indians on the frontier, and had resigned from the United States Army to accept service under his native State. Promoted from Colonel of the Forty-fifth Virginia Regiment to the command of a Virginia Brigade, he won additional promotion by his services in the Chancellorsville campaign (Spring of 1863), and was now at the head of a command ever to bear his name and to serve under him until he, with its shattered remnants, surrendered at Appomattox. "His earnest praise of the great qualities of his North Carolina soldiers was unstinted. Even to the last, there was a peculiar tension and quiver of the mouth when he would speak of their almost God-like heroism at Gettysburg, and the unheard of and never equalled slaughter that checked, but never terrified them."

MARCH TO GETTYSBURG.

Says a member of the regiment: "What a fine appearance the regiment made as it marched out from its bivouac near Fredericksburg that beautiful June morning. The men beaming in their splendid uniforms; the colors flying, and the drums beating; everything seemed propitious of success. On this march it was a real pleasure to see with what joy the people who had hitherto been under the domination of the Federals, received us. We marched by way of Harper's Ferry, where the gallows on which the notorious John Brown was hanged, was pointed out to us. Our Colonel was one of the cadets at the Virginia Military Institute at the time, and one of those who had guarded John Brown while awaiting his execution.

We crossed the Potomac at Shepherdstown and continued our march and rested beyond the little town of Fayetteville, Pa., on Sunday, 28 June, 1863. At this place the Chaplains held services.

Alas, the last Sunday on earth to many a noble soul then beating with such high hopes and aspirations. At this place some of the men of our brigade robbed a farmer of a few of his bee hives. This was regretted, for strict orders had been given that on this great march into the enemy's country, nothing should be taken except such provisions as the commissaries might require to be issued as rations and for which they were willing to pay. It being suggested that some of the men of the Twenty-sixth got some of the honey, Colonel Burgwyn and Lieutenant-Colonel Lane sought out the owner and paid him for it. The farmers along our line of march were quietly reaping and housing their grain. They did not seem to be in the least frightened or dismayed by our presence, and were left by us in the quiet and undisturbed possession of their crops.

On 30 June, we halted at a little village named Cashtown, on the Chambersburg Turnpike, about nine miles from Gettysburg, and were mustered preparatory to payment, and later in the afternoon proceeded to within about three and one-half miles of Gettysburg, just this side of a little creek, crossed by a stone bridge, where we filed to the right and bivouacked in a beautiful grove. That night Lieutenant-Colonel Lane was entrusted with the charge of the picket lines. After the establishment of the line, two ladies, much distressed and alarmed, because they were cut off from their houses, approached Colonel Lane who, assuring them that the Confederate soldier did not make war upon women and children, but ever esteemed it his duty and privilege to protect them, advanced the picket line beyond their homes, which lay close by.

The same day General Pettigrew, with three regiments of

his brigade, kept on to Gettysburg to procure shoes and other army supplies for his men; but meeting a strong force of the enemy's cavalry (two brigades of Buford's Division), and instructed not to bring on an engagement, General Pettigrew retraced his steps and rejoined the rest of the division in bivouac on the Chambersburg Turnpike, about three and a half miles distant from the village of Gettysburg. That night the men of Heth's Division quietly dreamed of home and loved ones in blissful ignorance of the momentous fact that Meade's great army was almost within their hearing.

GETTYSBURG, 1-3 JULY, 1863.

A warning carbine shot from a vidette of Buford's Cavalry Division on the bridge over Marsh Creek, fired in the early misty morn at the head of a column of infantry marching rapidly down the Chambersburg Turnpike, was the opening of the battle of Gettysburg This infantry column was the head of Heth's Division, marching to "feel the enemy" of whose presence the skirmish of the afternoon before, had apprised them. At once the leading brigade (Archer's) was filed to the right, formed in line of battle, its left resting on the turnpike and advanced to the front. Davis' brigade, forming in a similar manner on the left of the pike, with its right resting on the pike, also advanced. Pettigrew's and Brockenborough's Brigades, for the present, were held in reserve. Says a member of the Twenty-sixth Regiment: "As the head of the Twenty-sixth Regiment reaches the summit of the hill beyond the bridge crossing Marsh Creek, the enemy opens fire, sweeping the road with their artillery. There is some little excitement, but it soon disappears as Colonel Burgwyn riding along the line in his grandest style, commands in his clear, firm voice, 'Steady boys, steady.' "

The regiment filed off to the right about a hundred yards, when General Pettigrew and staff appeared on the field. He was mounted on his beautiful dappled gray. Never before had he appeared to greater advantage. His command was "echelon by battalion, the Twenty-sixth Regiment by the left flank." Colonel Burgwyn gave his Regiment the command, March! Then, as each regiment of the brigade marching to the right, uncovered the regiment in its front, its commander gave the order "By the left flank, March," and thus in a few moments, and by the quickest tactical movement the brigade was in line of battle, marching to the front in the following order from left to right, Twenty-sixth Regiment, Eleventh Regiment, Forty-seventh Regiment, and Fifty-second Regiment, each under the command of its respective Colonel.

Advancing in line of battle, the brigade was halted to await orders. Let us turn now to see what the Federals were doing.

On the night of 30 June, 1863, General Buford, in command of the advance division of cavalry of the Federal army, bivouacked his division on the western side of McPherson's ridge, which slopes down by a gentle descent to Willoughby's Run at the bottom. This ridge ran north and south, and about 400 yards to the west of the Seminary, which is about one-quarter of a mile to the west of Gettysburg. About 11 a. m. on 30 June, General Buford had entered Gettysburg by the Emmetsburg road, just as the head of Pettigrew's brigade was coming up on the Chambersburg turnpike, and as heretofore stated, there was a skirmish, and General Pettigrew withdrew, not wishing to bring on an engagement. At 10:30 that night, General Buford telegraphed General Meade "he is satisfied that A. P. Hill's Corps is massed just back of Cashtown." As Archer's Brigade advanced, it met Buford's pickets stretching along Willoughby run. Driving them in and rapidly advancing across the run, he struck Buford's main line—Gamble's Brigade composed of the Eighth New York, Eighth Illinois, two squadrons Twelfth Illinois, three squadrons Third Indiana Cavalry and Calif's Horse Artillery of six 3-inch rifle guns, now dismounted and acting as infantry, and posted along McPherson's ridge and in McPherson's woods. These troops Archer was steadily driving back up the slope, when he suddenly found himself enveloped between the extended lines of Meredith's (Iron) Brigade, of Wadsworth's Division of the First Army Corps just arrived on the scene at double quick. Major-General A. Doubleday in his report of the battle of Gettysburg, thus describes this action.

"The enemy (Archer's Brigade) were already in the woods and advancing at double quick to seize this central important position (McPherson's woods). The Iron Brigade led by the Second Wisconsin, in line followed by the other regiments, deployed en echelon, and without a moment's hesitation charged with the utmost steadiness and fury and hurled the enemy back into the run, and captured, after a sharp and desperate conflict, nearly one thousand prisoners, including General Archer. (General Heth places the number captured at 60 or 70.) General Archer was captured by Private Patrick Maloney, Company G, of the Second Wisconsin. Maloney was subsequently killed." "On the left," says General Heth, "Davis' Brigade advanced driving the enemy and capturing his batteries, but was unable to hold the position, the enemy concentrating on his front and flank an overwhelming force. The Brigade held its position until every field officer save two was shot down." By reference to General Wadsworth's report, it is seen that it was Cutler's Brigade, assisted by Second Maine Battery that was attacked by Davis' Brigade. General Wadsworth says: "The right became sharply engaged before the line was formed. At this time, 10:15 a. m., our gallant leader (General John F. Reynolds, commanding the First Corps, Army of the Potomac) fell mortally wounded. The regiments encountered heavy force, were outnumbered, outflanked and after a resolute contest, fell back in good order to Seminary Ridge near town. As they fell back, followed by the enemy, the Fourteenth New York State Militia, Sixth Wisconsin and Ninety-fifth New York Volunteers, gallantly charged on the advancing enemy and captured a large number of prisoners, including two entire regiments with their flags." Lieutenant-Colonel Rufus R. Dawes, commanding the Sixth Wisconsin, says in his report: "Major John A. Blair, commanding the Second Mississippi Volunteers, upon my demand, surrendered his sword and regiment to me, 7 officers and 225 men."

From this severe round, to use a pugilist's expression, both sides took a breathing spell and reformed to renew the attack. Says General Heth: "The enemy had now been felt and the division now was formed in line of battle on the right of the road as follows. Archer's, now commanded by Colonel B. D. Fry, of the Thirteenth Alabama, on the right; Pettigrew in the centre, and Brockenborough on the left. Davis' Brigade was kept on the left of the road to collect its stragglers; from its shattered condition it was not deemed advisable to bring it into action again on that day." It did, however, participate later in the action. After resting in line for an hour or more, orders came to attack the enemy in my front with the notification that Pender's Division would support me." Let us glance a moment at the character, numbers and position of the enemy which General Heth was now to assault with his two sound and one crippled brigade, and make, considering the fierceness with which it was made, the obstinacy with which it was met and the fearful loss in killed and wounded sustained on both sides, the most notable charge in all the battles of the war between the States.

A recent writer, John M. Vanderslice, author of a work called "Gettysburg. Then and Now," a gallant Union soldier, places the relative positions of the opposing forces at 11 a. m., 1 July, 1863, as follows: Heth's division occupied the extreme right, with Archer's Brigade on the right; next Pettigrew's, then Brockenborough's, then Davis'. Facing these Confederate troops, there was Meredith's Iron Brigade, occupying McPherson's woods. On the left of the woods was placed Biddle's Brigade and on the right of the woods was Stone's Brigade. The One Hundred and Fifty-first Pennsylvania Regiment of Biddle's Brigade was in reserve, so there were three regiments of that Brigade with Cooper's Battery in the action at the beginning. These several brigades were organized as follows: Meredith's Iron Brigade, Nineteenth Indiana, Twenty-fourth Michigan, Second, Sixth and Seventh Wisconsin Regiments

Biddle's Brigade, Eightieth New York, One Hundred and Twenty-first, One Hundred and Forty-second and One Hundred and Fifty-first Pennsylvania Regiments.

Stone's Brigade, One Hundred and Forty-third, One Hundred and Forty-ninth and One Hundred and Fiftieth Pennsylvania Regiments.

These regiments in these brigades were posted as follows:

Counting from left to right. Biddle's extreme left regiment One Hundred and Twentieth Pennsylvania. Next on right Eightieth New York, then Cooper's Battery, then One Hundred and Forty-second Pennsylvania. Meredith's Iron Brigade, extreme left regiment Nineteenth Indiana; next Twenty-fourth Michigan, next Seventh Wisconsin, and on the extreme right Second Wisconsin. The Sixth Wisconsin was in reserve. Stone's Brigade was not engaged with any of Pettigrew's men, but confronted the remnants of Davis' Brigade and the Forty-seventh and Fifty-fifth Virginia Regiments of Brockenborough's. Archer's Brigade on the Confederate extreme right overlapped Biddle's Brigade on the Federal extreme left, but Pettigrew's Brigade of four regiments, being in full ranks, and Biddle's three regiments not large, the two left regiments of Pettigrew's lapped over and confronted the left of the Iron Brigade, bringing the Twenty-sixth North Carolina Regiment with its 800 muskets in front of the Nineteenth Indiana and the Twenty-fourth Michigan, numbering together 784, rank and file.

The position of the Iron Brigade in McPherson's woods was not a straight line; the Nineteenth Indiana and Twenty-fourth Michigan formed nearly a straight line parallel with Willoughby Run, but its next regiment, the Seventh Wisconsin, on the right of the Twenty-fourth Michigan, was formed obliquely to the rear to confront an enemy attacking from its right flank, and also so as not to get outside of the protection of the woods, which General Doubleday says in his report "possessed all the advantages of a redoubt." Then on the right of the Seventh Wisconsin, the Second Wisconsin was formed connecting with the left of Stone's Brigade. Thus it appears the Twenty-sixth North Carolina regiment faced the front of the Iron Brigade, which consisted of the two regiments, the Nineteenth Indiana and the Twenty-fourth Michigan, but the Confederate troops charging these two regiments in the woods were subjected to the fire from the men of Biddle's Brigade and of Cooper's battery on their right; and it was from the fire of this battery, one of the best batteries of the Federal forces, that the Twenty-sixth regiment suffered severely, especially while charging across Willoughby Run, and reforming thereafter.

The situation then at 2 o'clock p. m., 1 July, 1863, is this: The Iron Brigade in line of battle in McPherson's woods is waiting the assault of Pettigrew's brigade, with the Twenty-sixth North Carolina Regiment of said brigade directly in their front, separated by Willoughby Run and disant about 300 yards.

The regiments of Pettigrew's Brigade were in line by echelon, the Twenty-sixth being in the advance and the Eleventh on its right some distance in the rear; the Forty-seventh regiment in rear of the Eleventh, and the Fifty-second in rear of the Forty-seventh. This made the Confederate troops appear to the enemy's vision, as in several lines of battle, whereas there was only one line of battle, and as the fight progressed, these regiments came up successively and formed one single line in the attack. They had, however, as their support Pender's division, some distance in the rear.

THE IRON BRIGADE.—ARMY OF THE POTOMAC.

The author of the History of the Twenty-fourth Michigan Regiment of this Brigade, thus accounts for its name and gives its record. Its cognomen, "Iron Brigade," was given them by General McClellan for intrepidity in the battle of South Mountain, 15 September, 1862. In proportion to its numbers it sustained the heaviest loss of any brigade in the Union army. Its loss at Gettysburg, first day's fight, was 1,153 out of 1,883 engaged, or 61 per cent. The Second Wisconsin sustained the greatest percentage of loss in killed and wounded of all the 2,000 regiments in the Union army. Its loss at Gettysburg was 77 per cent. of those engaged.

The Sixth Wisconsin had a total loss of 867 killed and wounded during the war, and the officer in command of the Second Mississippi Regiment of Davis' Brigade with 232 of his regiment and its colors, surrendered to this regiment in the early part of the first day's fight.

The Seventh Wisconsin met with the greatest loss of any regiment in the Union army at the battles of the Wilderness, and had 1,016 men killed and wounded during the war.

The Nineteenth Indiana in its first battle at Manassas, sustained a loss of 61 per cent., 259 out of 423 engaged, and the Twenty-fourth Michigan sustained the greatest loss of any regiment in the Union army at Gettysburg, 80 per cent., viz. 397 out of 496.

M'PHERSON'S WOODS.

General Doubleday says: "On the most westerly of these ridges (McPherson's) General Reynolds had directed his line to be formed. A small piece of woods (in the shape of a rectangular parallelogram) cut the line of battle in about two equal parts. These woods possessed all the advantage of a redoubt strengthening the centre of the line and enfilading the enemy's columns should they advance in the open spaces on either side. I deemed the extremity of the woods which extended to the summit of the ridge, to be the key of the position, and urged that portion of Meredith's (Iron) Brigade—the western men assigned to its defense—to hold it to the last extremity. Full of the memory of their past achievements, they replied cheerfully and proudly: 'If we can't hold it, where will you find the men who can?' "

Major John T. Jones, of the Twenty-sixth North Carolina Regiment, who commanded Pettigrew's Brigade after the third day's fight, and made the official report for the brigade, dated 9 August, 1863, thus describes the field:

"In our front was a wheat field about a fourth of a mile wide, then came a branch (Willoughby Run) with thick underbrush and briers skirting the banks. Beyond this again was an open field with the exception of a wooded hill (McPherson's woods) directly in front of the Twenty-sixth Regiment, and about covering its front. Skirmishers being thrown out, we remained in line of battle until 2 p. m., when orders to advance were given."

THE CHARGE.

The Twenty-sixth was the extreme left regiment of Pettigrew's Brigade. It directly faced McPherson's woods and its front about covered the width of the woods. The Iron Brigade occupied these woods; the open space on the left of the woods (Confederate right) was defended by Biddle's Pennsylvania Brigade of four regiments with Cooper's Battery in the centre, the open space on the right of the woods (Confederate left) was defended by Stone's Pennsylvania Brigade with three regiments. Stewart's Battery B, Fourth United States Artillery attached to the Iron Brigade, was posted on the right and rear supporting Stone's Brigade, but in a position to sweep any part of the field. A Northern writer says: "There is no doubt, more men fell at Stewart's guns than in any other battery in the Union armies." Company F, of the Twenty-sixth Regiment, was on the left of the colors. Company E on the right and Companies A and G near the centre. The position of these companies nearest the flag accounts for their disproportionate losses in the battle.

A member of the Twenty-sixth regiment thus describes the situation: "While we were still lying down impatiently waiting to begin the engagement, the right of the regiment was greatly annoyed by some sharpshooters stationed on the top of a large old farm house to our right. Colonel Burgwyn ordered a man sent forward to take them down, when Lieutenant J. A. Lowe, of Company G, volunteered. Creeping forward along a fence until he got a position from whence he could see the men behind the chimney who were doing the shooting, he soon silenced them.

During all this time, Hill was bringing up his Corps and placing it in position. Colonel Burgwyn became quite impatient to engage the enemy, saying we were losing precious time; but Hill did not come, and we had nothing to do but to wait for his arrival on the field. However, we were keeping our men as quiet and comfortable as possible, sending details to the rear for water, and watching the movements of the enemy. The enemy's sharpshooters occasionally reminded us that we had better cling close to the bosom of old mother earth.

Many words of encouragement were spoken and some jokes were indulged in. Religious services were not held, as they should have been, owing to the absence of our Chaplains. All this time the enemy were moving with great rapidity. Directly in our front across the wheat field was a wooded hill

(McPherson's woods). On this hill the enemy placed what we were afterwards informed was their famous "Iron Brigade." They wore tall, bell-crowned black hats, which made them conspicuous in the line. The sun was now high in the heavens. General Ewell's Corps had come up on our left and had engaged the enemy. Never was a grander sight beheld. The lines extended more than a mile, all distinctly visible to us. When the battle waxed hot, now one of the armies would be driven, now the other, while neither seemed to gain any advantage. The roar of artillery, the crack of musketry and the shouts of the combatants, added grandeur and solemnity to the scene. Suddenly there came down the line the long awaited command "Attention." The time of this command could not have been more inopportune; for our line had inspected the enemy and we well knew the desperateness of the charge we were to make; but with the greatest quickness the regiment obeyed. All to a man were at once up and ready, every officer at his post, Colonel Burgwyn in the center, Lieutenant-Colonel Lane on the right, Major Jones on the left. Our gallant standard-bearer, J. B. Mansfield, at once stepped to his position—four paces to the front, and the eight color guards to their proper places. At the command "Forward, march!" all to a man stepped off, apparently as willingly and as proudly as if they were on review. The enemy at once opened fire, killing and wounding some, but their aim was too high to be very effective. All kept the step and made as pretty and perfect a line as regiment ever made, every man endeavoring to keep dressed on the colors. We opened fire on the enemy. On, on, we went, our men yet in perfect line, until we reached the branch (Willoughby's Run) in the ravine. Here the briers, reeds and underbrush made it difficult to pass, and there was some crowding in the centre, and the enemy's artillery (Cooper's Battery) on our right, getting an enfilade fire upon us, our loss was frightful; but our men crossed in good order and immediately were in proper position again, and up the hill we went, firing now with better execution.

The engagement was becoming desperate. It seemed that the bullets were as thick as hail stones in a storm. At his post on the right of the regiment and ignorant as to what was taking place on the left, Lieutenant-Colonel Lane hurries to the center. He is met by Colonel Burgwyn, who informs him "it is all right in the centre and on the left; we have broken the first line of the enemy," and the reply comes, "we are in line on the right, Colonel."

At this time the colors have been cut down ten times, the color guard all killed or wounded. We have now struck the second line of the enemy where the fighting is the fiercest and the killing the deadliest. Suddenly Captain W. W. McCreery, Assistant Inspector General of the Brigade, rushes forward and speaks to Colonel Burgwyn. He bears him a message. "Tell him," says General Pettigrew, "his regiment has covered itself with glory today." Delivering these encouraging words of his commander, Captain McCreery, who had always contended that the Twenty-sixth would fight better than any regiment in the brigade, seizes the flag, waves it aloft and advancing to the front, is shot through the heart and falls, bathing the flag in his life's blood. Lieutenant George Wilcox, of Company H, now rushes forward, and pulling the flag from under the dead hero, advances with it. In a few steps he also falls with two wounds in his body.

The lines hesitates; the crisis is reached; the colors must advance. Telling Lieutenant-Colonel Lane of the words of praise from their brigade commander just heard, with orders to impart it to the men for their encouragement, Colonel Burgwyn seizes the flag from the nerveless grasp of the gallant Wilcox, and advances, giving the order "Dress on the colors." Private Frank Honeycutt, of Company B, rushes from the ranks and asks the honor to advance the flag. Turning to hand the colors to this brave young soldier, Colonel Burgwyn is hit by a ball on the left side, which, passing through both lungs, the force of it turns him around and, falling, he is caught in the folds of the flag and carries it with him to the ground. The daring Honeycut survives his Colonel but a moment and shot through the head, now for the thirteenth time the regimental colors are in the dust.

Kneeling by his side, Lieutenant-Colonel Lane stops for a moment to ask: "My dear Colonel, are you severely hurt?" A bowed head and motion to the left side and a pressure of the hand is the only response; but "he looked as pleasantly as if victory was on his brow." Reluctantly leaving his dying commander to go where duty calls him, Lieutenant-Colonel Lane hastens to the right, meets Captain McLauchlin, of Company K, tells him of General Pettigrew's words of praise, but not of his Colonel's fall; gives the order "Close your men quickly to the left. I am going to give them the bayonet"; hurries to the left, he gives a similar order, and returning to the center finds the colors still down. Colonel Burgwyn and the brave boy private, Franklin Honeycut, lying by them. Colonel Lane raises the colors. Lieutenant Blair, Company I, rushes out, saying: "No man can take these colors and live." Lane replies: "It is my time to take them now"; and advancing with the flag, shouts at the top of his voice: "Twenty-sixth, follow me." The men answer with a yell and press forward. Several lines of the enemy have given away, but a most formidable line yet remains, which seems determined to hold its position. Volleys of musketry are fast thinning out those left and only a skeleton line now remains. To add to the horrors of the scene, the battle smoke has settled down over the combatants making it almost as dark as night. With a cheer the men obey the command to advance, and rush on and upward to the summit of the hill, when the last line of the enemy gives way and sullenly retires from the field through the village of Gettysburg to the heights beyond the cemetery.

Just as the last shots are firing, a sergeant in the Twenty-fourth Michigan Regiment (now the President of the Iron Brigade Veteran Association, Mr. Charles H. McConnell, of Chicago), attracted by the commanding figure of Colonel Lane carrying the colors, lingers to take a farewell shot, and resting his musket on a tree, he waits his opportunity. When about thirty steps distant, as Colonel Lane turns to see if his regiment is following him, a ball fired by this brave and resolute adversary, strikes him in the back of the neck just below the brain, which crashes through his jaw and mouth, and for the fourteenth and last time the colors are down. The red field was won, but at what a cost to the victor as well as to the vanquished.

LOSSES IN THE FIRST DAY'S FIGHT.

Pettigrew's brigade was opposed on the first day at Gettysburg to the best troops in the Federal army, viz: Biddle's Pennsylvania and Meredith's (Iron) Brigade of Western troops. The Twenty-sixth Regiment fought at one or another period of the charge, the Nineteenth Indiana and Twenty-fourth Michigan, of the Iron Brigade, and the One Hundred and Fifty-first Pennsylvania, of Biddle's Brigade, which came to the support of the Federal second line. Says the author of "Gettysburg, Then and Now," published in 1899: "While the fighting had been going on upon the Federal right Pettigrew also made a desperate attack on Biddle's Brigade. The Fifty-second North Carolina overlapping the line had attacked the One Hundred and Twenty-first Pennsylvania on the left flank, compelling it to change front and the Forty-seventh and Eleventh North Carolina encountered the Twentieth New York and One Hundred and Forty-second Pennsylvania, while at the same time the Twenty-sixth North Carolina fighting its way up the woods, was penetrating a gap between the One Hundred and Forty-second Pennsylvania and the Nineteenth Indiana, of Meredith's (Iron) Brigade, the left of which had been forced back.

At this juncture the One Hundred and Fifty-first Pennsylvania which was in reserve near the Seminary, rushed to the front and met the Twenty-sixth North Carolina in one of the bloodiest struggles that took place on the field, as will be noticed when the losses of these regiments are stated."

Quoting again from Major Jones' official report of the part taken by Pettigrew's Brigade in the battle of Gettysburg, he says:

"The Brigade moved forward in beautiful style, in quick time, on a line with the brigade on our left commanded by Colonel Brockenborough. When nearing the branch (Willoughby Run) the enemy poured a galling fire into the left of the brigade from the opposite bank where they had massed in heavy force, while we were in line of battle awaiting the order to advance. The Forty-seventh and Fifty-second North Carolina, although exposed to a hot fire from artillery and infantry, lost but few men in comparison with the Eleventh and Twenty-sixth. On went the command across the branch and up the opposite slope, driving the enemy at the point of the bayonet back upon their second line.

"The second line was encountered by the Twenty-sixth regiment, while the other regiments were exposed to a heavy artillery shelling. The enemy's single line in the field on our right, was engaged principally with the right of the Eleventh North Carolina and the Forty-seventh North Carolina. The enemy did not perceive the Fifty-second North Carolina, which flanked their left until the Fifty-second discovered themselves by a raking and destructive fire by which the enemy's line was broken.

"On the second line the fighting was terrible, our men advancing, the enemy stubbornly resisting, until the two lines were pouring volleys into each other at a distance not greater than twenty paces. At last the enemy were compelled to give way. They again made a stand in the woods, and the third time they were driven from their positions losing a stand of colors which was taken by the Twenty-sixth regiment, but owing to some carelessness, they were left behind and were picked up by some one else."

Let us quote now from the other side in obedience to the maxim *"Fas est ab hoste doceri."* Colonel Henry A. Morrow, Twenty-fourth Michigan, a native of Warrenton, Va., who as a young man moved to Detroit, Mich., and was a City Judge there in 1862, and raised the regiment of which he was appointed to the command, in his report of the battle, says: "I gave directions to the men to withhold their fire until the enemy should come within short range of our guns. This was done. Their advance was not checked and they came on with rapid strides yelling like demons. The Nineteenth Indiana, on our left, fought most gallantly, but was forced back. The left of my regiment was now exposed to an enfilade fire and orders were given for this portion of the line to swing back so as to face the enemy now on our flank. Pending the execution of this movement, the enemy compelled me to fall back and take a new position a short distance in the rear.

"The second line was promptly formed and we made a desperate resistance, but we were forced to fall back and take up a third position beyond a slight ravine. My third color-bearer was killed on this line. Augustus Ernst, Company K.

"By this time the ranks were so diminished that scarcely a fourth of the force taken into action could be rallied. Captain Andrew Wagner, Company F, one of the color guard, took the colors and was ordered by me to plant them in a position to which I designed to rally the men. He was wounded in the breast and left the field. I now took the flag from the ground where it had fallen and was rallying the remnant of my regiment, when Private William Kelly, of Company E, took the colors from my hands, remarking as he did so, 'The Colonel of the Twenty-fourth Michigan shall never carry the colors while I am alive.' He was killed instantly. Private Lilburn A. Spaulding, Company K, seized the colors and bore them for a time. Subsequently I took them from him to rally the men and kept them until I was wounded.

"We had inflicted severe loss on the enemy, but we were unable to maintain our position, and were forced back step by step, contesting every foot of the ground to the barricade west of the Seminary building. The field over which we fought from our first line of battle in McPherson's woods to the barricade near the Seminary, was strewn with the killed and wounded.

"Our losses were very large, exceeding perhaps the losses sustained by any regiment of equal size in a single engagement of this or any other war. The strength of the regiment on 1 July, 1863, was 28 officers and 468 rank and file, total 496. We lost, killed 8 officers and 59 men. Wounded, 13 officers and 197 men. Missing or captured, 3 officers and 83 men. Nearly all our wounded, myself among them, fell into the hands of the enemy. The flag of the regiment was carried by no less than nine persons, four of the number were killed and three wounded. All the color guard were killed or wounded."

Returning to Confederate sources for accounts of the heroic conduct of the Twenty-sixth North Carolina Regiment, I quote from his official report of the battle, made by Major-General Heth, commanding the division:

"Pettigrew's Brigade under the leadership of that gallant officer and accomplished scholar, Brigadier-General J. Johnston Pettigrew (now lost to his country), fought as well and displayed as heroic courage, as it was ever my fortune to witness on a battlefield. The number of its own gallant dead and wounded as well as the large number of the enemy's dead and wounded left on the field over which it fought, attests better than any communication of mine, the gallant part it played on 1 July. In one instance, when the Twenty-sixth North Carolina Regiment encountered the second line of the enemy, its (Twenty-sixth Regiment's) dead marked its line of battle with the accuracy of a line at dress parade."

Under date of 9 July, 1863, less than a week before his fatal wounding at Falling Waters (14 July, 1863), General Pettigrew writes Governor Vance as follows: "Knowing that you would be anxious to hear from your old regiment, the Twenty-sixth, I embrace an opportunity to write you a hasty note. It covered itself with glory. It fell to the lot of the Twenty-sixth to charge one of the strongest positions possible. They drove three, and we have every reason to believe, five regiments out of the woods with a gallantry unsurpassed. Their loss has been heavy, very heavy, but the missing are on the battlefield and in the hospital. Both on the first and third days your old command did honor to your association with them and to the State they represent."

Captain J. J. Young, regimental Quartermaster of the Twenty-sixth regiment, under date of 4 July, 1863, writes Governor Vance as follows:

"The heaviest conflict of the war has taken place in this vicinity. It commenced July 1st, and raged furiously until late last night. Heth's Division, A. P. Hill's Corps, opened the ball and Pettigrew's Brigade was the advance. We went in with over 800 men in the regiment. There came out of the first day's fight 216 all told, unhurt. Yesterday they were again engaged, and now have only about 80 men for duty. To give you an idea of the frightful loss in officers,

Heth being wounded, Pettigrew commanded the division (Pettigrew had the bones of his left hand crushed by a grape shot, but remained on the field with his hand in splints), and Major Jones our brigade. (Jones was also slightly wounded, but refused to leave the field). Eleven men were shot down the first day with the colors (afterwards ascertained to be fourteen). Yesterday they were lost. Poor Colonel Burgwyn was shot through both lungs and died shortly afterward. His loss is great, for he had few equals of his age. Captain W. W. McCreery, Inspector on General Pettigrew's staff, was shot through the heart and instantly killed. Assistant Adjutant-General N. Collins Hughes mortally wounded. Lieutenant Walter M. Robertson, Brigade Ordnance Officer, severely wounded; with them, Lieutenant-Colonel Lane through the neck, jaw and mouth, I fear mortally; Adjutant James B. Jordan in the hip, severely; Captain J. T. Adams, shoulder, seriously; Stokes McRae, thigh broken; Captain William Wilson, killed; Lieutenants W. W. Richardson and J. B. Holloway have died of their wounds. It is thought Lieutenant M. McLeod and Captain N. G. Bradford will die; Captain J. A. Jarrett, wounded in face and hand. Yesterday Captain S. P. Wagg was shot through by grape, and instantly killed. Alex. Saunders was wounded and J. R. Emerson left on the field dead. Captain H. C. Albright is the only Captain left in the regiment. Lieutenants J. A. Lowe, M. B. Blair, T. J. Cureton (this officer was wounded in shoulder), and C. M. Sudderth are the only officers not wounded. Major Jones was struck by a fragment of a shell on the 1st and knocked down and stunned on the 3rd, but refused to leave the field.

"Our whole division numbers only 1,500 or 1,600 effective men as officially reported, but, of course, a good many will still come in. The division at the beginning numbered about 8,000 effective men. Yesterday in falling back we had to leave the wounded, hence the uncertainty of a good many being killed yesterday evening."

Going into particulars of losses: Company F, from Caldwell County, commanded by Captain R. M. Tuttle (now a Presbyterian minister at Collierstown, Va.), went into the battle with three officers and 88 muskets. Thirty-one were killed or died of wounds received in the battle. Sixty were wounded, fifty-nine of whom were disabled for duty. Sergeant Robert Hudspeth was the only man able to report for duty after the fight, and he had been knocked down and stunned by the explosion of a shell. In this company were three sets of twin brothers, at the close of the battle, five of the six lay dead on the field.

Companies I and F of this regiment were from Caldwell County. The men composing these companies had been reared along the slopes of the Great Grandfather Mountain. They had been accustomed from boyhood to hunt deer, the bear, and the wolf in the lonely forests surrounding their homes. They were enured to hardship, self-reliant, indefatigable and insensible to danger. Company F was on the left of the colors, and Company E on the right. This latter company (Company E) suffered nearly as badly as Company F. It carried 82 officers and men into the fight, and brought out only two untouched.

Going into the particulars of the loss of Company E, 18 men were killed or mortally wounded, and 52 wounded on the first day, and on the third day only two escaped. Every officer in the company was wounded.

Company G lost 12 men killed and 58 wounded and missing, but the losses on each day are not given by Captain Albright.

Company H had 17 killed and 55 wounded in the two days battles.

The men composing these three companies were from the historic counties of Chatham and Moore. Their ancestors had fought at Alamance and Moore's Bridge and Guilford Court House, and from their youth up they had handled the rifle in hunting the deer and wild turkey, and as General Pettigrew said of them, "they shot as if they were shooting at squirrels."

Company A, from Ashe County, the same class of mountaineers of whom we have spoken above in referring to Companies F and I, took into action 92, rank and file. Eleven were killed and 66 wounded in the first day's fight, and on the third day, its Captain (Wagg) was killed, and ten wounded and missing out of fourteen taken into the fight. Lieutenant J. A. Polk, commanding Company K when the muster roll was signed 31 August, 1863, states every officer was wounded at Gettysburg, 16 men killed and 50 wounded and missing. He does not give the number taken into action.

As to the loss sustained by the regiment as a whole, we may rely upon the statements of Northern writers who have compiled them from the official records in the War Department at Washington, D. C. Colonel William F. Fox, of Albany, N. Y., in his book, "Regimental Losses in the Civil War," a work of recognized authority—places the loss of the Twenty-sixth Regiment in the first day's fight at 86 killed and 502 wounded, out of 800 taken into action. He says: "On the third day's fight in Pickett's charge, they lost 120, recorded as missing." In a letter to the writer dated 30 September, 1889, Colonel Fox says: "My figures for the loss of the Twenty-sixth North Carolina at Gettysburg, are taken from the official report of Surgeon-General Lafayette Guild, C. S. A., who obtained his figures from the nominal lists of the killed and wounded made out in the field hospitals. In my opinion, the 120 missing should also be included in the killed and wounded; but as they were not so reported officially, I cannot substitute my opinion for official statistics. In a second edition, which is now going through the press, I added the losses for Bristoe Station, having obtained them from the War Department since the publication of the first edition. In these losses for Bristoe, I was surprised to see that the Twenty-sixth North Carolina again heads the list. I took great pains to verify the loss of the Twenty-sixth North Carolina at Gettysburg, for I am inclined to believe that in time this regiment will become as well known in history as the Light Brigade at Balaklava."

Colonel Fox further states in his book that this loss of the Twenty-sixth Regiment was the greatest in numbers and greatest in per cent. of those taken into action of all the regiments on either side in the Civil War in any one battle. Mr. John M. Vanderslice, Director of the Gettysburg Memorial Association, who was a private in Company D, Eighth Pennsylvania, was gazetted for distinguished services in action at

Hatcher's Run, 6 February, 1865, in his book, "Gettysburg, Then and Now"—writes thus: "The loss of the Twenty-sixth North Carolina Regiment should be 584 on the first day and of the remaining 216, 130 were lost on the third, its total loss in the battle being 588 killed and wounded and 126 missing out of 800 engaged. This brigade (Pettigrews's) lost over 500 additional on the third day."

As a matter of historical interest, I append a list of the losses in the several brigades that fought in and around McPherson's woods on the first day at Gettysburg:

		Killed and Wounded.	Missing.	Engaged.
Union Troops.	Meredith s Iron Brigade—			
	2 Wisconsin	182	51	302
	6 Wisonsin	146	22	
	7 Wisconsin	126	52	402
	19 Indiana	160	50	338
	24 Michigan	272	91	496
	Biddle's Brigade—			
	80 New York	146	24	287
	121 Pennsylvania	118	61	263
	142 Pennsylvania	141	70	291
	151 Pennsylvania	262	75	467
	Stone's Brigade	574	279	
	Artillery	105		
	Gamble's Cavalry	83	28	
Heth's Division.	Davis' Mississippi Brigade	695		
	Archer's Tennessee Brigade	160		
	Brockenborough's Virginia Brigade	148		
	Pettigrew's North Carolina Brigade—			
	11 North Carolina Regiment	209		
	26 North Carolina Regiment	588		
	47 North Carolina Regiment	161		
	52 North Carolina Regiment	147		
		1105		

THIRD DAY'S BATTLE AT GETTYSBURG, 3 JULY, 1863.

Quoting again from Major John T. Jones' report: "The night of the first day's fight (1 July, 1863) the brigade bivouacked in the woods they had occupied previously to making the charge. We remained in this position until the evening of the 2nd, when we moved about a mile to our right and took position in rear of our batteries facing the works of the enemy on Cemetery Hill. We remained here until about 12 o'clock on the 3rd, when our batteries opened upon the enemy's works. About 2 o'clock we were ordered to advance."

A member of the regiment thus writes:

"On the second day, Pettigrew's entire brigade rested. General Pettigrew showed great energy in recruiting his thinned ranks. He commanded that all those not too severely wounded should return to active duty and armed all the cooks and extra duty men and every other man in any way connected with the regiment. The regimental band (Captain Mickey's band) was ordered to play inspiring music to cheer the soldiers, whose spirits were depressed at the loss of so many of their comrades, and in every way the condition of things was enlivened. On the evening of the 2nd, General Pettigrew marched his command to the place in the line from which the grand charge was to be made next day. To the great surprise of every one, the brigade seemed as ready for the fray on the morning of the third day, as it had been on that of the first."

PICKETT'S AND PETTIGREW'S CHARGE.

Quoting from the author of "Gettysburg, Then and Now": "There were two hours of comparative silence until 1 o'clock p. m. when the signal gun was fired from Seminary Ridge, by the Washington Artillery of New Orleans, and there was opened between the 138 Confederate and the 80 Federal guns the heaviest and most terrible artillery fire ever witnessed upon any battle field in this country, if upon any in the world. It opened so suddenly that the men were torn to pieces before they could rise from the ground upon which they had been lolling. Some were stricken down with cigars in their mouths. One young soldier was killed with the portrait of his sister in his hand. The earth was thrown up in clouds. Splinters flew from fences and rocks, and mingled with the roar of the artillery were the groans of wounded men and the fierce neighing of mangled horses.

"In the meantime the fresh troops of Pickett's Confederate division had been massed under cover of the slight ridge running between Seminary Ridge and the Emmettsburg road in rear of the artillery. While Pettigrew's Division (formerly Heth's) was massed to their rear and left behind Seminary Ridge. In the rear of the right of Pickett were the brigades of Wilcox and Perry, with that of Wright in reserve.

"In the rear of the right of Pettigrew were the brigades of Scales, and Lane, of Pender's Division, commanded by Trimble. When the artillery ceased firing, these troops moved from behind their cover and advanced majestically across the field towards Cemetery Hill. Pickett's Division on the right, Pettigrew's on its left and rear en echelon, supported by Scales' and Lane's brigades. Pickett's division was in line as follows: Kemper's Brigade on the right, Garnett on his left, while Armistead was in the rear. On the left of Pickett were the four brigades of Pettigrew's division. Archer's Brigade, commanded by Frye, next to Pickett; Pettigrew's, commanded by Marshall, of the Fifty-second North Carolina Regiment, next on the left; Davis next, and Brockenborough on the extreme left.

"In the rear of Frye and Marshall, there were Scales' Brigade, commanded by Lowrance, and Lane's Brigade, these under Major-General Trimble, from Maryland. Together the assaulting columns numbered 14,000. The point of direction was the small copse of trees to the left of Ziegler's Grove, held by Gibbon's Division of the Second Corps. After advancing some distance the three brigades of Pickett's division made a half wheel to the left in order to move toward the objective point. McGilvery's forty guns (Federal artillery) on the left, with those of the two batteries on Round Top, opened a terrible fire upon them. As the division neared the wall, it was joined on its left by Frye's Brigade, and at the same time Lowrance's North Carolina Brigade rushed from its rear and joined Frye's and Garnett at the angle of the wall. The two guns of Cushing's battery at the wall were silenced.

"The left of that charging column under Pettigrew and Trimble, suffered as severely as the right under Pickett. Great injustice has been done these troops by the prevailing erroneous impressions that they failed to advance with those of Pickett.

"Such is not the fact, as they were formed behind Seminary Ridge they had over 1,300 yards to march under the terrible fire to which they were exposed, while Pickett's Division being formed under cover of the intermediate ridge, had but 900 yards to march under fire. At first, the assaulting columns advanced en echelon, but when they reached the Emmettsburg road, they were on a line, and together they crossed the road. The left of Pettigrew's command becoming first exposed to the fearful enfilading fire upon their left flank from the Eighth Ohio, and other regiments of Hay's Division and of Woodruff's Battery and other troops, the men on that part of the line (Brockenborough's Brigade) either broke to the rear or threw themselves on the ground for protection.

"But Pettigrew's other brigades, Davis, Marshall and Frye, with the brigades of Lowrance and Lane, under Trimble, advanced with Pickett to the stone wall and there fought desperately. As the assaulting column reached the wall, Wilcox's Alabama and Perry's Florida Brigade to the right, marching according to order, but becoming separated from Pickett, had resumed the march to the left, and were now advancing from the top of the crest, from behind which Pickett had emerged, directly towards McGilvery's batteries and the Third Corps, but received by a severe fire from Stannard's Vermonters, who had changed front again, and exposed to a severe artillery fire and seeing the commands of Pickett, Pettigrew and Trimble repulsed, they withdrew under cover of the hill. Thus ended this reckless and ever renowned effort to carry Cemetery Hill by direct assault in the face of 100 cannon and the Federal Army."

Quoting from Major Jones' report, he says:

"About 2 o'clock we were ordered to advance. It was an open field about three-quarters of a mile in width. In moving off there was some confusion in the line, owing to the fact that it had been ordered to close in on the right on Pickett's division, while that command gave way to the left. This was soon corrected, and the advance was made in perfect order. When about half across the intermediate space the enemy opened on us a most destructive fire of grape and canister. When within about 250 or 300 yards of the stone wall behind which the enemy was posted, we were met by a perfect hail storm of lead from their small arms. The brigade dashed on and many had reached the wall when we received a deadly volley from the left. The whole line on the left had given way, and we were being rapidly flanked, and with our thinned ranks and in such a position it would have been folly to stand against such odds.

"After this day's fight but one field officer was left in the brigade, and regiments that went in with Colonels came out commanded by Lieutenants."

A member of the Twenty-sixth Regiment thus describes the charge:

"As soon as the fire of the artillery ceased, General Pettigrew, his face lit up with the bright look it always wore when in battle, rode up to Colonel Marshall, in command of the brigade, and said: 'Now Colonel, for the honor of the good Old North State. Forward.' Colonel Marshall promptly repeated the command, which taken up by the regimental commanders, the *Twenty-sixth* marched down the hill into the valley between the two lines. As the forward march continued, our artillery would occasionally fire a shot over the heads of the troops to assure them that they had friends in the rear.

"The brigade had not advanced far when the noble Marshall fell, and the command of the brigade devolved on Major Jones, of the Twenty-sixth, while that of the regiment on Captain S. W. Brewer, of Company E, a man who proved on that day as he has often since, that he was thoroughly qualified to lead.

"The Confederate line was yet unbroken and still perfect, when about half a mile from their works the enemy's artillery opened fire, sweeping the field with grape and canister; but the line crossed the lane (Emmettsburg road) in good order. When about 300 yards from their works the musketry of the enemy opened on us, but nothing daunted the brave men of the Twenty-sixth pressed quickly forward and when the regiment reached within about forty yards of the enemy's works, it had been reduced to a skirmish line. But the brave remnant still pressed ahead and the colors were triumphantly planted on the works by J. M. Brooks and Daniel Thomas, of Company E, when a cry came from the left, and it was seen that the entire left of the line had been swept away.

"The Twenty-sixth now exposed to a front and enfilade fire, there was no alternative but to retreat, and the order was accordingly given. Captain Cureton, of Company B, and others, attempted to form the shattered remnants of the regiment in the lane (Emmettsburg road) but pressed by the enemy, the attempt was abandoned.

General Pettigrew had his horse shot under him during the charge, and though wounded (bones of his left hand shattered by a grape shot) he was one of the last men of his division to leave, and was assisted off the field by Captain Cureton, whom he ordered to rally and form Heth's division behind the guns for their support. Pettigrew's brigade promptly responded and formed when told where to go.

"By night a very good skirmish line had been collected and the gallant old Twenty-sixth had 67 privates and 3 officers present on the night of 3 July, 1863, out of 800 who went into battle on the morning of 1 July. In this enumeration the cooks and extra duty men and others who had been armed are not counted. These 70 officers and men remained to support the artillery that night and all next day."

As of historical interest, I append the losses of Pickett's, Pettigrew's and Trimble's Division on this third day's fight at Gettysburg.

	Killed and Wounded	Missing.
Pickett's Division—		
Garnett's Brigads, 8, 18, 19, 28 and 56 Virginia Regts.....	402	539
Armistead's Brigade, 9, 14, 38, 53 and 57 Virginia Regts...	574	643
Kemper's Brigade, 1, 3, 7, 11 and 24 Virginia Regts.......	462	317
	1438	1499
Pettigrew's Division—		
Archer's Brigade..	330	112
Pettigrew's Brigade	300	228
Davis' Brigade.........	244	160
	874	500
Trimble's Division—		
Lane's Brigade.............	264	176
Scales' Brigade	125	85
	389	261

Adding the killed and wounded of Pettigrew's Brigade on the third day's fight, viz., 300; to its killed and wounded on the first day's fight, viz., 1,105; and it makes a total loss of 1,405 killed and wounded sustained by these four North Carolina Regiments, which is within 33 of the loss in killed and wounded sustained by the fifteen Virginia Regiments of Pickett's Division.

PICKETT OR PETTIGREW.

Quoting again from the author of "Gettysburg, Then and Now": "But why call this Pickett's charge? In this assault there were engaged forty-two Confederate Regiments. In Pickett's Division there were 15 Virginia Regiments. In Pettigrew's and Trimble's there were 15 North Carolina Regiments, 3 Mississippi, 3 Tennessee, 2 Alabama and 4 Virginia Regiments. In addition to the artillery fire, they (Pettigrew and Trimble) encountered 9 Regiments of New York, 5 of Pennsylvania, 3 of Massachusetts, 3 of Vermont, 1 Michigan, 1 Maine, 1 Minnesota, 1 New Jersey, 1 Connecticut, 1 Ohio, and 1 Delaware, in all 27 regiments.

"Some prominent writers, even historians like Swinton and Lossing, have said that the left of the line (Pettigrew's and Trimble's Divisions) did not advance as was expected, and that it was because these troops were not of the same 'fine quality' as those upon the right; that they were raw and undisciplined, etc., etc. Yet, but two days before, these same soldiers of Pettigrew and Trimble had fought around Reynold's Grove (McPherson's woods) for six hours in a struggle with the First Corps that is unsurpassed for bravery and endurance, and where so many of their numbers had fallen. There were in fact no better troops in the Confederacy than they. Is history repeating herself? If the event is correctly recorded, there were at Thermopylæ 300 Spartans, 700 Thespians, and 300 Thebans. It is said the latter went over to the enemy, but the Thespians died to a man at the pass with the Spartans. Yet for nearly twenty-four centuries, Epic song and story have well preserved the memory of the Spartans, while the devoted Thespians are forgotten."

INCIDENTS OF THE BATTLE.

On the first day while the Twenty-sixth Regiment was in line awaiting the order to charge the enemy in McPherson's woods, Lieutenant-Colonel Lane, who had been up all the night previous in charge of the division skirmish line,and had eaten but little, but had drunken freely of muddy water, was seized with an intolerable nausea and vomiting. Colonel Lane thus speaks of the incident: "I asked permission of Colonel Burgwyn to go to the rear. The latter replied: 'Oh, Colonel, I can't, I can't, I can't think of going into this battle without you; here is a little of the best French brandy which my parents gave me to take with me in the battle; it may do you good.' I took a little of it under the circumstances, though I had not drunk any during the war, and I may add, neither had Colonel Burgwyn. In a few minutes I was somewhat relieved and said: 'Colonel Burgwyn, I can go with you.' With his usual politeness, he replied: 'Thank you, Colonel, thank you.' Continuing the conversation, he said: 'Colonel, do you think that we will have to advance on the enemy as they are? Oh, what a splendid place for artillery. Why don't they fire on them?' He saw and realized the very decided advantage their position gave them over us."

James D. Moore, private in Company F, was the 85th man of his company shot on the first day's fight. A ball passed through his leg. When taken to the field hospital the surgeon said he had been fighting cavalry, as the wound was made by a carbine 44 calibre, and not by an Enfield rifle, 56-calibre. After the war Moore went to live in Indiana at a place called Winnamac. He there met a man named Hayes who was a member of the Twenty-fourth Michigan Regiment and in the battle of Gettysburg. Hayes had lost his Enfield rifle on the forced march of the night before, and as his regiment was going into action on the morning of 1 July, he picked up a carbine dropped by one of Buford's cavalry, and used it during the fight. It was the only carbine in the Twenty-fourth Regiment and just before he retreated, when the colors of the regiment charging him was fifteen or twenty paces distant, he fired in their direction. Moore at the time was alongside the flag and received Hayes' shot. They became good friends and Hayes was of material assistance to Moore so long as the latter lived in his town.

When taken from the field, Colonel Lane was carried to the field hospital, a brick house. A wounded Georgia officer, who was lying near the door of the room in which Colonel Lane was, had been delirious all the morning. He finally became quiet about 1 p. m. and after a silence of some minutes, Colonel Lane heard him say in a perfectly rational tone of voice: "There now, there now. Vicksburg has fallen, General Lee is retreating and the South is whipped. The South is whipped." He ceased speaking and in a few moments an attendant passed by and said he was dead. General Lee did not retreat from Gettysburg until the evening of the 4th of July, and Vicksburg was not surrendered until the 4th of July.

It is stated in Volume 67, page 514, Official Records Union and Confederate Armies, that on 4 July, 1863, at 6:35 a. m., General Lee proposed to General Meade "to promote the comfort and convenience of the officers and men captured by the opposing armies, that an exchange be made at once." At 8:25 a. m., General Meade replied: "It is not in my power to accede to the proposed arrangement."

COLONEL LANE ESCAPES CAPTURE.

When the army retreated from Gettysburg, the wounded were sent off in long trains chiefly of the wagons which General Stuart had captured in his raid around Meade's army. These invited the attack of the enemy's cavalry, and many wounded Confederate officers and soldiers were in this way captured before the army got across the Potomac river.

The wagon train in which Colonel Lane was carried, was one of those attacked. He at once got out of the wagon, mounted his horse and made his escape, though he was at the time unable to speak or to receive nourishment in the natural way. He was unable to take any nourishment for nine days, owing to the swollen and inflamed condition of his throat and mouth, and it was thought impossible for him ever to get well.

OFFICERS PRESENT AT THE BATTLE.

Posterity will wish to know as much as possible of the personnel of this regiment, and we append a list of the officers of the regiment who participated in the battle of Gettysburg. This we are enabled to do from a very remarkable fact.

As stated above, the proximity of Meade's army was not known on 30 June, 1863, and on that day the regiment was mustered as it bivouacked after the day's march. These muster and pay rolls were made out in triplicate, one to be sent to the Adjutant General of the army, one to be kept by the company commander, and one by the Quartermaster of the regiment, who was also the paymaster. Captain J. J. Young, the regimental Quartermaster from the beginning to the end of the war, has preserved these muster and pay rolls. The writer has had access to the same, and now copies from them the names of the officers of the regiment who were present in camp on the afternoon of 30 June, 1863, and the number of

TWENTY-SIXTH REGIMENT.

1. John Tuttle, Sergeant, Co. F.
2. Wm. N. Snelling, 2d Lieut., Co. D.
3. L. L. Polk, Sergeant Major
4. W. W. Edwards, Private, Co. E.
5. J. D. Moore, Private, Co. F. (The 85th man in his Company wounded at Gettysburg, July 1st, 1863.)
6. H. C. Coffey, Private, Co. F. (The 86th man in his Company wounded at Gettysburg, July 1st, 1863.)
7. Laban Ellis, Private, Co. E.

those present for duty in each company as shown by its muster and pay roll for that day.

FIELD AND STAFF.

Harry King Burgwyn, Jr., Colonel.
John Randolph Lane, Lieutenant-Colonel.
John Thomas Jones, Major.
James B. Jordan, Adjutant.
Llewellyn P. Warren, Surgeon.
William W. Gaither, Assistant Surgeon.
Joseph J. Young, Quartermaster.
Phineas Horton, Commissary.
Montford S. McRea, Sergeant Major.
Benjamin Hind, Hospital Steward.
Abram J. Lane, Quartermaster Sergeant.
Jesse F. Ferguson, Commissary Sergeant.
E. H. Hornaday, Ordnance Sergeant.

COMPANY OFFICERS PRESENT.

Company A—Captain, Samuel P. Wagg; First Lieutenant, A. B. Duvall; Second Lieutenant, J. B. Houck; Junior Second Lieutenant, L. C. Gentry; present for duty, 97.

Company B—Captain, Wm. Wilson; First Lieutenant, Thos. J. Cureton; Second Lieutenant, W. W. Richardson; Junior Second Lieutenant, Edward A. Breitz; present for duty, 92.

Company C—Captain J. A. Jarrett; First Lieutenant, W. Porter; Second Lieutenant, ———; Junior Second Lieutenant, R. D. Horton; present for duty, 80.

Company D—Captain, J. T. Adams; First Lieutenant, Gaston Broughton; Second Lieutenant, J. G. M. Jones; Junior Second Lieutenant, Orren A. Hanner; present for duty, 79.

Company E—Captain, S. W. Brewer; First Lieutenant, John R. Emerson; Second Lieutenant, W. J. Lambert; Junior Second Lieutenant, Oran A. Hanner; present for duty, 104.

Company F—Captain, R. M. Tuttle; First Lieutenant, C.

M. Sudderth; Second Lieutenant, — — ———; Junior Second Lieutenant, J. B. Holloway; present for duty, 91.

Company G—Captain, H. C. Albright; First Lieutenant, J. A. Lowe; Second Lieutenant, — — ———; Junior Second Lieutenant, Wm. G. Lane; present for duty, 91.

Company H—Captain, — — ———; First Lieutenant, M. McLeod; Second Lieutenant, George Willcox; Junior Second Lieutenant, J. H. McGilvery; present for duty, 78.

Company I—Captain, N. G. Bradford; First Lieutenant, M. B. Blair; Second Lieutenant, J. C. Grier; Junior Second Lieutenant, J. G. Sudderth; present for duty, 74.

Company K—Captain, James C. McLauchlin; First Lieutenant, Thomas Lilly; Second Lieutenant, — — ———; Junior Second Lieutenant, J. L. Henry; present for duty, 99.

The total number present for duty was 885.

Of those absent, Captain James D. McIver of Company H, Second Lieutenant A. B. Hays of Company F, and Second Lieutenant A. R. Jordan of Company G, were absent on detached duty, Second Lieutenant Wm. L. Ingram of Company K, was on sick furlough, and Second Lieutenant J. M. Harris of Company C, who was subsequently captured at Bristoe Station (14 October, 1863) is marked "absent with leave."

Of the above list those killed or mortally wounded in the two days' fighting, were as follows: Colonel, H. K. Burgwyn; Captains S. P. Wagg, Wm. Wilson; Lieutenants, John R. Emerson, W. W. Richardson, J. B. Holloway.

All the other officers except Captain Albright and Lieutenants J. A. Lowe, C. M. Sudderth and M. B. Blair, were wounded. Adjutant J. B. Jordan and Sergeant-Major M. S. McRea, of the Regimental Staff, both severely wounded. Major Jones and Lieutenant T. J. Cureton were wounded, but refused to leave the field.

WOUNDED OFFICERS CAPTURED.

Captains, Bradford and Brewer. Lieutenants, Brietz, Broughton, Hanner, McLeod, and Adjutant Jordan.

On 31 August, 1863, while the regiment was in camp near Orange Court House, it was again mustered. The writer has these rolls before him. In some companies the record of events since 30 June, 1863 (last muster) is specific; in some, no details are given other than what appears opposite the name of the individual.

Captain Duval, of Company A, reports that his company went into action at Gettysburg with 92 men and lost, killed 11, and wounded 66, on the first day, and on the third day, 1 killed, Captain Wagg, and 10 wounded and missing; total, 88.

First Lieutenant W. J. Lambert, of Company E, says his company took into the battle 82 men and lost, killed and mortally wounded 18, and wounded 52, on the first day, and on the second day's fight only two men escaped.

Captain Albright, of Company G, reports the loss of his company at 12 killed and 58 wounded and missing.

Captain McIver, of Company H, reports 17 killed and 55 wounded at Gettysburg.

Lieutenant Polk, of Company K, says he recrossed the Potomac at Falling Waters with 16 men, having crossed that river in June on the way to Gettysburg, with 103, rank and file.

Captain Tuttle, of Company F, declares that every man was killed or wounded in his company that he took into the battle.

The following is the number killed and wounded and missing at Gettysburg, ascertained from the reports as given on the muster rolls of the companies, dated 31 August, 1863: "Killed and mortally wounded, 139. Wounded and missing, 535."

This enumeration omits some wounded who had returned to duty prior to 31 August, 1863, the date of the muster.

The muster rolls for 30 June, 1863, make the aggregate present for duty, enlisted men, 885; allowing 10 per cent. for extra duty and details, it would leave about 800 muskets taken into battle at Gettysburg on the first day. Of this number 708 were killed, wounded and missing as the losses in the first and third day's fighting at Gettysburg. Over 88 per cent—and of the officers, 34 out of 39 were killed or wounded. Over 87 per cent.

****

COLOR BEARERS AT GETTYSBURG.

It is possible at this late day that the name of some gallant soldier who carried the flag of the Twenty-sixth Regiment during the battle of Gettysburg may be omitted from the list below, but every effort has been made to include in this honorable mention all entitled, for no one took the flag in that battle without the certainty of being shot down, and not one escaped.

The color guard consisted of a Sergeant and eight privates. After these nine had fallen, the others were volunteers.

FIRST DAY'S FIGHT, 1 JULY, 1863.

Colonel, H. K. Burgwyn, Jr., killed.
Captain Wm. W. McCreery, killed.
Private Franklin Honeycutt, Company B, killed.
" John R. Marley, Company G, killed.
" William Ingram, Company K, killed.
Lieutenant-Colonel John R. Lane, wounded.
Lieutenant George Willcox, wounded.
Color Sergeant J. Mansfield, wounded.
Sergeant Hiram Johnson, Company G, wounded.
Private John Stamper, Company A, wounded.
" G. W. Kelly, Company D, wounded.
" L. A. Thomas, Company F, wounded.
" John Vinson, Company G, wounded.

THIRD DAY'S FIGHT, 3 JULY, 1863.

Sergeant W. H. Smith, Company K, killed.
Private Thomas J. Cozart, Company F, killed.
Captain S. W. Brewer, Company E, wounded.
Private Daniel Thomas, Company E, wounded.

As First Sergeant James M. Brooks, Company E, and Daniel Thomas, the latter carrying the flag, reached the enemy's works, the Federals called out to them, "Come over on this side of the Lord," and took them prisoners rather than fire at them.

LITTER BEARERS AT GETTYSBURG.

These men kept right up with the regiment. I have only been able to locate the following names:

Private Neill B. Staton, Company B.
" Jackson Baker, Company D.
" John A. Jackson, Company H.

FALLING WATERS—DEATH OF GENERAL PETTIGREW.

On the night of 4 July, 1863, General Lee withdrew his army from confronting Meade at Gettysburg, and Heth's Division marched to Hagerstown, where it entrenched. "On the evening of 13 July," says General Heth in his official report, "I received orders to withdraw at dark and move in the direction of Falling Waters. The night was dark, roads ankle deep in mud and raining. It took twelve hours to march seven miles. On reaching an elevated and commanding ridge of hills, one mile from Falling Waters, I was ordered by General A. P. Hill to put my division in line of battle on either side of the road and to put Pender's Division in rear of mine in column of brigades. At this point we halted to let the wagons and artillery get over the river. About 11 a. m. 14 July, 1863, received orders to move Pender's division across the river following Anderson's Division. About 15 or 20 minutes after getting these orders, and while they were in execution, a small body of cavalry, numbering 40 or 45, made their appearance in our front. They were at once observed by myself and General Pettigrew, and several members of my staff as well as many others. On emerging from the woods the party faced about, apparently on the defensive. Suddenly facing about, they galloped up the road and halted some 175 yards from my line of battle. From their manœuvering and the smallness of their numbers, I concluded it was a party of our own cavalry pursued by the enemy. In this opinion I was sustained by all present. The troops had been restrained up to this time from firing by General Pettigrew and myself. Examining them critically with my glasses, I discovered they were Federal troops, and the command was given to fire. At the same time the Federal officer gave the command to charge. The squad passed through the intervals separating the epaulments for the artillery and fired several shots. In less than three minutes all were killed or captured, save two or three who are said to have escaped. General Pettigrew, who had received a wound in one of his hands (left) at Gettysburg, was unable to manage his horse which reared and fell with him. It is probable when in the act of rising that he was struck by a pistol ball in the left side, which, unfortunately for himself and his country, proved fatal. Thirty-three of the enemy's dead were counted, and six prisoners fell into our hands and a stand of colors."

The cavalry mentioned above was a portion of the Sixth Michigan, commanded by Major P. A. Weber. "Seeing only that portion of the enemy behind the earthworks," says General Kilpatrick in his report of the affair, "Major Weber gave the order to charge."

General Kilpatrick admits a loss of thirty killed, wounded and missing, including the "gallant Major P. A. Weber, killed." It would seem that General Heth and the rest were excusable for their hesitation as to which side this cavalry force belonged. 'Tis true, they were dressed in the Federal uniform, but many Confederate scouts wore the Federal uniform. It was known that General Lee was crossing his army into Virginia, at Williamsport ford and at Falling Waters on a pontoon bridge, and that the cavalry had orders to protect the crossing of the infantry at these places. But for an unfortunate mistake on the cavalry's part in thinking all had crossed, whereby those who were to intervene between the enemy and Heth's rear guard had been withdrawn and had, themselves, crossed at Williamsport above, this sad disaster could not have occurred.

A member of the Twenty-sixth regiment, who witnessed the unfortunate affair says: "Some (referring to the Federal cavalry) were knocked off their horses with fence rails. General Pettigrew after he fell, endeavored to shoot the Yankee who shot him, but his pistol missed fire, and N. B. Staton, private of Company B, seized a big stone and crushed the Yankee in the breast, killing him."

As soon as the surgeons examined General Pettigrew's

wound they saw the only hope for his life was to keep him perfectly quiet, and proposed to take him into a barn near by. To allow this, General Pettigrew obstinately declined, saying "he would die before he would again be taken prisoner." He was then put on a stretcher, and in hopes his life by this way might be saved, he was carried by four men who were regularly relieved by fresh details, all the way to Bunker Hill, a distance of 22 miles, occupying parts of two days. Frequently during the march he would say to the soldiers as he would notice their sympathetic countenances: "Boys, don't be disheartened. May be I will fool the doctors yet." He lingered in the house of a Mr. Boyd, at Bunker Hill, Va., until 17 July, 1863, and at about half past six in the morning, died quietly and without pain. General Lee, riding by his side as he was carried on the litter to Bunker Hill, expressed great sorrow at his being wounded. General Pettigrew replied "that his fate was no other than one might reasonably anticipate upon entering the army, and that he was perfectly willing to die for his country."

To the Rev. Mr. Wilmer, afterwards Bishop Wilmer, of Louisiana, he avowed a firm persuasion of the truths of the Christian religion and said that in accordance with his belief he had, some years before, made preparation for death.

On the morning of Friday, 24 July, 1863, the coffin containing his remains, wrapped in the flag of his country, and hidden under wreaths of flowers and other tributes of feminine taste and tenderness, lay in the rotunda of the Capitol at Raleigh, where within the year had preceded him his compatriots, Branch and Anderson. From Raleigh, he was taken to his old home, Bonarva, Lake Scuppernong, Tyrrell County, and there he is buried near the beautiful lake whose sandy shores his youthful feet were wont to tread. We would pause here to remark how mysterious are the dispensations of Providence, that it should be denied to James Johnston Pettigrew to die on the field of Gettysburg, and be decreed that he must meet his end in a petty skirmish with cavalry two weeks later.

Many prisoners were taken on the retreat from Hagerstown to Falling Waters, because of the exhausted condition of the men and the incessant pursuit of the Federal cavalry.

The troops at Falling Waters had to cross a pontoon bridge. The Confederate cavalry having retreated across at Williamsport, there were none to protect the infantry of Heth's division as it crossed at Falling Waters. The enemy's cavalry pressed them on front and flank, and there was more or less demoralization at the last.

Captain Cureton, of Company B, witnessed this incident. A Federal cavalryman took position near the Maryland end of the pontoon bridge and as the stragglers came along he would demand their surrender. In this way some fifty or sixty men had surrendered to this one cavalryman, when a member of the Twenty-sixth Regiment passing along, was halted and his surrender demanded. The Twenty-sixth Regiment man raised his gun and taking aim said: "Damn you, you surrender." The Federal said "all right," and threw down his gun. He was taken prisoner and with the fifty or sixty who had surrendered to him, was marchel across the bridge by the Tar Heel. Captain Cureton was the last man to get on the pontoon bridge as it was cut loose from its Maryland end and swung into the river. From a thousand to fifteen hundred stragglers were left on the Maryland side by this premature cutting loose of the bridge, and fell into the enemy's hands.

BRISTOE STATION, 14 OCTOBER, 1863.

After the return to Virginia from the Gettysburg campaign, General Lee stationed his army in and around Orange Court House. While here on 7 September, 1863, General Wm. W. Kirkland was appointed to the command of Pettigrew's Brigade, and remained in command until the battle of Bristoe Station, where he was wounded.

In a letter from General Lee to President Davis, dated 17 October, 1863, he thus describes this unfortunate engagement: "With a view of bringing on an engagement with the army of General Meade, this army marched on the 9th instant by way of Madison Court House and arrived near Culpepper on the 11th. The enemy retired towards the Rappahannock. We only succeeded in coming up with a portion of his rear guard at this place (Bristoe Station) on the 14th instant, when a severe engagement ensued, but without any decided or satisfactory results."

In his eagerness to attack the retiring enemy (Third Army Corps) General A. P. Hill overlooked the presence of the Second Corps posted behind the railroad embankment in a cut; and when the brigades of Cooke and Kirkland made the attack, they were suddenly confronted by the Second Corps posted as above stated, and were driven back with severe loss. In his report of the engagement, General A. P. Hill says: "In conclusion I am convinced I made the attack too hastily; at the same time a delay of half an hour and there would have been no enemy to attack. In that event I believe I should equally have blamed myself for not attacking at once."

The losses sustained by Kirkland's brigade in this action:

Regiment.	Killed.	Wounded.
Eleventh	4	11
Twenty-sixth	16	83
Forty-fourth	23	63
Forty-seventh	5	37
Fifty-second	2	25
Total	50	219

WINTER OF 1863-'64.—THE SNOW BALL BATTLE.

The Army of Northern Virginia winter-quartered in and around Orange Court house.

"At the first heavy fall of snow, it was suggested that there should be a sham battle between Cooke's and Kirkland's Brigades, and snow balls be the weapons used. The two brigades paraded facing each other on opposite sides of a ravine. Colonel Wm. MacRae, of the Fifteenth North Carolina Regiment, commanded Cooke's Brigade; as to the name of the commander of Kirkland's, the writer is not advised. At a given signal the battle began. At first the men contented themselves with using snow, and all was fun and frolic; but as the contest waxed more animated and each side struggled for mastery, the passions of the combatants became aroused and the excitement of actual battle seized them; hard substances, frequently stones, were grabbed up with the snow and made into a ball that had the stinging effect of the genuine article on the one hit, and several received injuries of a serious nature. Colonel MacRae was pulled from his horse and roughly handled, and the combat only ended with the exhaustion of the participants, each side agreeing it should be considered a drawn battle. This affair caused some bitterness between the brigades, which took time and comradeship, battles, privation and sufferings to destroy."

About the middle of November, 1863, Colonel Lane having sixty days longer leave of absence, visited his regiment. He thus writes of his visit: "I found the regiment so low in spirits and few in number that the day I reached camp, was, I believe, the saddest day to me of all the war. I realized then, as not before, the deaths of my Colonel, Harry Burgwyn, of our General, Pettigrew, and so many other officers and friends in the regiment.

"Regretting so much to see the gallant old regiment go down, notwithstanding the fact that I was entirely unable for active service, I reported myself for duty, when I was commissioned as full Colonel of the Twenty-sixth Regiment, to date from 1 July, 1863. I went to work with all the will I could possibly bring to bear to recruit, drill and equip my regiment and restore it to something like its former numbers and efficiency."

Major John T. Jones had been promoted Lieutenant-Colonel, after the battle of Gettysburg, and at one time commanding the brigade, had been in command of the regiment from Gettysburg until Colonel Lane's return. Captain Jas. T. Adams, of Company D, on his return to the regiment after his recovery from his wound received at Gettysburg (first day) was promoted to Major. The commissions of all bearing date 1 July, 1863, in recognition of the heroic conduct of the regiment on that day. The captaincy of Company D was held open awaiting the return of First Lieutenant Gaston H. Broughton, wounded and captured in the third day's fight at Gettysburg. Orderly Sergeant John A. Polk, of Company K, promoted Second Lieutenant after Gettysburg, where he was wounded, was appointed acting Adjutant, vice

Adjutant Jordan, wounded and captured at Gettysburg.

Continuing our quotations from Colonel Lane's letter: "I was informed by General Kirkland that if consolidation of regiments were effected, that the Twenty-sixth Regiment was named as one to be consolidated. I used every influence at my command to avert the threatened consolidation, and through the noble concert of action of the officers of the regiment, I had the proud satisfaction of seeing our efforts crowned with success.

"Such was the harmony, energy and regimental pride of the officers and men, and so well did they work together to promote its interests, enlivened by such soul-inspiring music as only Captain Mickey's band could furnish, that by the first of May, 1864, the regiment numbered 760 strong; and so well was it drilled that General Heth pronounced it to be one of the 'best drilled regiments in the Army of Northern Virginia.' The improvement in the moral and religious condition of the regiment that winter was very remarkable, more good being effected by the work of the Chaplains and their assistants than during all the previous years of the war."

Many deserters returned, gave themselves up and ever afterwards made good soldiers, and by 5 May, 1864, this old Twenty-sixth Regiment that had been bereft of so many of its best officers and men at Gettysburg, and Bristoe Station, that it came near losing its separate existence by being merged into another, proudly marched down the plank road at the head of Heth's division to the

BATTLES OF THE WILDERNESS AND SPOTTSYLVANIA COURT HOUSE.

On 4 May, 1864, General U. S. Grant, now in command of the armies of the United States, with General Meade in immediate command of the Army of the Potomac, crossed the Rapidan at Ely and Germania fords.

General Lee marched two corps to oppose him. Ewell's (Second Corps) by the old turnpike, and Hill's (Third Corps) by the Orange plank road.

Says General Lee in his report of the battle: "Ewell and Hill arrived in the morning in close proximity to the enemy's line of march. A strong attack was made upon Ewell, who repulsed it, capturing many prisoners and four pieces of artillery. The enemy subsequently concentrated on Hill, who, with Heth's and Wilcox's Divisions, successfully resisted repeated and desperate assaults. Early on the morning of 6 May, as these divisions were being relieved, the enemy advanced and created some confusion. The ground lost was recovered so soon as the fresh troops got into position and the enemy were driven back. Afterward we turned the left of his front line and drove it from the field. Lieutenant-General Longstreet was severely wounded."

A member of the regiment thus writes of this battle:

"Never did a regiment march more proudly and determinedly than the Twenty-sixth, when it headed the column of Kirkland's Brigade for the battle of the Wilderness. Reaching the ground early 5 May, 1864, we passed General Lee and his Staff. Our regiment was engaged all the first day, and succeeded in driving back the enemy and holding him in check; but informed we would be relieved during the night by men of Longstreet's Corps, we did not take proper precaution and were surprised by the enemy, who at daybreak next morning (6 May) with great vigor, renewed the attack of the previous afternoon, and our brigade came very near being stampeded. And again the regiment met with serious loss in prisoners and killed and wounded."

Colonel Lane being wounded in the thigh on the evening before, Lieutenant-Colonel Jones was now in command of the regiment, and while gallantly rallying his men and leading them in a charge, was mortally wounded. He asked Assistant Surgeon W. W. Gaither, if the wound was mortal. When told it was, with a yearning expression he replied: "It must not be. I was born to accomplish more good than I have done." Later on will be found a sketch of this noble, gallant young soldier who died ere his prime, but left a proud record behind him. Continuing our quotation: "The regiment succeeded in holding the lines and at the critical moment, Longstreet came up with his magnificent corps in the most perfect order I ever saw, marching his forces against

Grant like boys going to a frolic. He hurled back the enemy and getting in their rear and left flank, was driving them in great confusion from the field, when, like Stonewall Jackson, General Longstreet fell, shot down by some of his own men (part of Mahone's Brigade) and the pursuit was stopped. After Lieutenant-Colonel Jones was wounded, Colonel Lane returned to duty, his wound not proving very severe.

"Lee and Grant now moved along on parallel lines fronting each other like two great monsters, and the night of 7 May, found Lee's army well in line, fronting Grant, with Longstreet's Corps, commanded by Anderson on the right, Ewell on the left, and Hill in the center, the Twenty-sixth Regiment being near the centre of Hill's Corps, placed it about the centre of the army.

THE REBEL YELL.

"About 8 p. m., on the night of 7 May, it became rumored that Grant's army was moving to his left, and had lost hope of reaching Richmond by the overland route. The rebel yell was raised at some point on the right of the line; at first, heard like the rumbling of a distant railroad train, it came rushing down the lines like the surging of the waves upon the ocean, increasing in loudness and grandeur; and passing, it would be heard dying away on the left in the distance. Again it was heard coming from the right to die away again on the distant left. It was renewed three times, each time with increased vigor. It was a yell like the defiant tones of the thunder storm, echoing and re-echoing. It caused such dismay among the Federals that it is said their pickets fired and ran in."

During the night General Lee put his army in motion for Spottsylvania Court House, and arrived just in time, as the enemy came in sight about 9 a. m. next morning (8 May).

The 10th was a day of vigorous battle, the enemy made incessant attacks on the First Corps (Andersons), but were continually repulsed with great slaughter. During the night of the 11th, the artillery protecting Johnstons Division at the salient was withdrawn to be ready to move to the right, when at dawn of the 12th, Hancock's Corps attacked and captured it, and most of Johnston's Division and twenty guns. It has been stated that Johnston was surprised by the enemy on this occasion. This he denies. In his report of the affair he says: "On the night of 11 May, in riding around my lines, I found the artillery leaving the trenches and moving to the rear. About 12 p. m. I communicated to Lieutenant-General Ewell my belief that I would be attacked and requested the return of the artillery. There was no surprise; my men were up and ready for the assault before the enemy made their appearance."

A member of the Twenty-sixth Regiment writes:

"At the battle of Spottsylvania Court House, the Twenty-sixth was detached from its regular place in the line and stationed about fifty yards from the Court House to be in readiness to support any point which might be strongly assaulted. While we were yet lying there, General Lee came riding by on his war horse, Traveler. Grant's artillery opened fire and it seemed impossible that General Lee could escape in the storm of shot and shell which was centered upon him. As quick as a flash the members of his staff placed themselves around him to protect him with their own bodies. Such was the sentiment in the entire army. Each one was willing to give up his life to save that of the Commander-in-Chief. The troops were visibly affected, as General Lee with his staff, still surrounding him, rode off. This incident manifested the love, reverence and respect in which General Lee was held by his soldiers."

At a critical time in the campaign it was extremely difficult to get corn for the artillery horses. Three farmers living a few miles up the river tendered General Lee two thousand bushels of corn, but the trouble was, how to get it, as it was necessary to send a wagon train for it and the road lay for a greater part of the distance in close proximity to the lines of the enemy. As an escort for this wagon train, General Lee ordered that some regiment should be selected to whose officers the men yielded unquestioned obedience, and upon whom they had entire reliance. The Twenty- sixth Regiment was selected for this hazardous service; the corn was safely brought into camp and the hungry artillery horses fed, making it possible to move the guns, and thus relieving the army from a threatened disaster.

BRIGADIER-GENERAL WM. MACRAE.

On his recovery from the wound received at Bristoe Station, General Kirkland was in command of the brigade until he was again wounded on 2 June, 1864, when Colonel Wm. MacRae, of the Fifteenth North Carolina Regiment, of Cooke's Brigade, was made Brigadier-General, and assigned to the command of Kirkland's Brigade 27 June, 1864. General MacRae is thus spoken of by officers of the regiment:

"General MacRae soon won the confidence and admiration of the brigade, both officers and men. His voice was like that of a woman; he was small in person, and quick in action. To him history has never done justice. He could place his command in position quicker and infuse more of his fighting qualities into his men, than any officer I ever saw. His presence with his troops seemed to dispel all fear, and to inspire every one with a desire for the fray. The brigade remained under his command until the surrender."

Another officer thus writes:

"General MacRae assigned to the brigade changed the physical expression of the whole command in less than two weeks, and gave the men infinite faith in him and themselves, which was never lost, not even when they grounded arms at Appomattox."

FROM THE WILDERNESS TO RICHMOND.

On all the line from the Wilderness to Richmond and Petersburg, General Lee acted on the defensive. He suffered the enemy to attack him, and in every instance the result proved the wisdom of his doing so. General Lee had not a man to lose unnecessarily. There were no reserves for him to call upon to fill his depleted ranks. Not so his adversary. As a matter of historical interest, I will quote briefly from some of General Grant's dispatches to General Halleck at Washington, D. C., giving the losses in his army on this march to Richmond:

"4 May, 1864: The crossing of the Rapidan effected. Forty-eight hours will now demonstrate whether the enemy intend giving battle this side of Richmond." It has been shown that in less than twelve hours from the date of this dispatch Lee had inflicted a severe repulse upon Grant's army.

"6 May, 11:30 a. m.: We have been engaged with the enemy in full force since early yesterday. I think all things are progressing favorably. Our loss to this time I do not think exceeds eight thousand.

"7 May, 10 a. m.: Our losses to this time in killed, wounded and prisoners will not exceed twelve thousand.

"11 May, 1864: We have lost up to this time, eleven general officers, killed, wounded and missing, and probably twenty thousand men.

"26 May, 1864: Lee's army is really whipped. The prisoners we now take show it, and the action of his army shows it unmistakably. A battle with them outside of their intrenchments cannot be had. Our men feel that they have gained the morale over the enemy and attack with confidence." A few days later, General Grant's tone is different.

"5 June, 1864: Without a greater sacrifice of human life than I am willing to make, all cannot be accomplished that I had designed. I have, therefore, resolved upon the following plan: Move to the south side of James river."

It is now well known that so disheartened was the army of the Potomac by its fearful losses in killed, wounded and missing from the crossing of the Rapidan to and including the battle of Cold Harbor, June 1-3, 1864 (the official reports make this loss over forty thousand), that at the latter battle the soldiers refused to obey the orders to attack the Confederate lines. (In this last battle the Federals lost over ten thousand), and General Grant in his testimony before the Congressional Committee investigating the cause of the failure at the Mine explosion (at Petersburg 30 July, 1864) gave it as one of the explanations for the failure, the detail of white troops rather than Ferrero's Division of negroes, to make the assault, the white troops being demoralized from their life in the trenches and losses in battle.

From Spottsylvania Court House to the North Anna, at Hanover Junction, Cold Harbor, on the lines between Richmond and Petersburg, the Twenty-sixth was always prompt to respond to all orders. General Grant, like Wm. Taylor's snake, would "wire in and wire out, and frequently left us still in doubt, whether he was coming in or going out."

INCIDENTS OF THIS CAMPAIGN.

On two occasions while on the picket line between Spottsylvania Court House and Richmond, Colonel Lane's life was probably saved by the vigilance of his men.

On one occasion Private Laban Ellis, of Company E, seeing a Federal soldier taking aim at the Colonel, fired so quick that his ball struck the Federal's gun as it went off and knocked it from his shoulder, whereupon the latter surrendered and said to Colonel Lane: "Your man saved you." On another occasion, as Colonel Lane, with Ira Nall, also of Company E, were making a reconnoissance of the ground in their front, Nall spied a man a few feet away with his gun leveled upon the Colonel. Without taking time to raise his gun to his shoulder, Nall fired and brought the Federal down, killing him.

It would be impossible to state in detail all the engagements in which the regiment participated along this line. General Grant attempted to go around us, over us, and under us (explosion of the mine, 30 June, 1864), but was foiled in every attempt. Two of the most brilliant victories in which MacRae's Brigade played a conspicuous part were the engagements at *Davis House, 19 August,* and *Reams Station, 25 August, 1864.* In General Lee's reports of these actions, he thus writes 20 August, 1864: "General Hill attacked the enemy (Fifth Corps) yesterday afternoon at Davis House, three miles from Petersburg, on Weldon Railroad, defeated him and captured about 2,700 prisoners, including one Brigadier-General, and several field officers."

26 August, 1864: "General A. P. Hill attacked the enemy in his entrenchments at Reams Station yesterday evening and at the second assault, carried his entire line. Cooke's, MacRae's and Lane's Brigades (under General Connor), and Pegram's artillery, composed the assaulting column. Hill captured nine pieces of artillery, twelve colors, 2,150 prisoners, 3,100 stand of small arms and 32 horses."

So altogether creditable was the conduct of these three North Carolina Brigades as to call forth from General Lee a letter to Governor Vance, dated 29 August, 1864, in which he says: "I have frequently been called upon to mention the services of the North Carolina soldiers in this army, but their gallantry and conduct were never more deserving of admiration than in the engagement at Reams Station, on the 25th instant. The brigades of Generals Cooke, MacRae and Lane, the last under the command of General Connor, advanced through a thick abatis of felled trees under a heavy fire of musketry and artillery and carried the enemy's works with a steady courage that elicited the warm commendation of their corps and division commanders, and the admiration of the army. If the men who remain in North Carolina share the spirit of those they have sent to the field, as I doubt not they do, her defense may be securely entrusted in their hands."

INCIDENTS IN THE BATTLE—MAJOR GENERAL HETH A JOINT COLOR BEARER.

The troops selected to carry the enemy's works in the early part of the fight having been repeatedly driven back, Heth's Division was ordered to their assistance. The division was drawn up in line of battle with the skirmishers in front.

Lieutenant D. C. Waddell, of Company G, Eleventh North Carolina Regiment, relates this incident to the writer. Lieutenant Waddell was in command of the skirmishers on that part of the line. Major-General Heth walked out to his line and ordered him to send a man back to the main line and bring a regimental flag. The messenger returned with the color-bearer of the Twenty-sixth Regiment. General Heth demanded the flag. The color-bearer refused to give it up, saying: "General, tell me where you want the flag to go and I will take it. I won't surrender up my colors." The General again made the demand, and was met by the same refusal, when taking the color-bearer by the arm, he said: "Come on then, we will carry the colors together." Then giving the signal to charge by waving the flag to the right and the left, the whole line with a yell, started for the enemy's works. The abatis protecting the enemy's lines was interlaced with wire in places, but charging through and over and around it all, the Confederate line rushed up to the works, and General Heth, and his co-color-bearer, planted the flag on the entrenchments behind which lay the enemy, most of whom thereupon surrendered. Thomas Minton, of Company C, from Wilkes County, was the name of this gallant color-bearer. He was subsequently killed with his colors in the action near Burgess Mill, 27 October, 1864. This gallant soldier was also wounded at Gettysburg.

This courageous assault was necessarily attended with considerable loss in killed and wounded. Coloned Lane was again so unfortunate as to be wounded. He was struck by a piece of shell in the left breast just over the heart, fracturing two ribs and breaking one and tearing open the flesh to the bones, making a fearful wound six inches long and three wide, from which it was thought he would surely die. But about the first of November he was again back with his command ready for duty.

Among the other officers of the Twenty-sixth Regiment killed in these almost daily engagements with the enemy, was Captain Henry C. Albright, of Company G. He fell mortally wounded at the head of his company in repulsing an attack on the Vaughn Roads, 29 September, 1864. It would seem he had a presentment of his death. Captain Albright had been in every engagement and battle in which his regiment participated from New Bern, up to that day, and escaped from even a slight wound. On the day he was wounded he remarked to a friend: "Oh, how I dread this day." He was carried to the Winder hospital, insisting that he be placed in the ward where his soldier boys were, rather than in the Officer's hospital. He lingered until 27 October, 1864. He

TWENTY-SIXTH REGIMENT.

1. J. D. McIver, Captain, Co. H.
2. Thomas Lilly, Captain, Co. K.
3. Jas. C. McLauchlin, Captain, Co. K
4. W. W. Gaither, Assistant Surgeon.
5. George Wilcox, 1st Lieut., Co. M.
6. Orran A. Hanner, 1st Lieut., Co. E.
7. J. G. Jones, 1st Lieut., Co. D.

was carried home and buried in his family grave yard at Pleasant Hill, Chatham County. A handsome monument marks the spot.

He was succeeded by First Lieutenant A. R. Johnson, who was such a martinet that the boys called him "Bob Ransom." Few companies in the Confederate army had better officers than Company G. Lieutenant-Colonel James T. Adams was now in command of the Twenty-sixth, and remained so until Colonel Lane returned to duty as stated above.

Heth's Division being on the extreme right of the Confederate line defending Petersburg, were among the troops first to be called upon to resist any flank movement on the part of General Grant; and there was fighting almost daily along their front and flank.

At Burgess Mills, 27 October, 1864, where Hancock lost 1,482 in killed and wounded; on Warren's expedition with the Fifth Corps to destroy Weldon bridge when he was met and driven back at Belfield 7-12 December, 1864; in the severe engagements at Hatcher's Run, 5-6 February, 1865, with Warren's Corps (Fifth) and Gregg's Division of cavalry, in which Warren admits a loss of 1,376 killed and wounded and missing; in all these actions MacRae's Brigade was actively engaged and maintained its high prestige to the end. Of the suffering borne without murmuring, and fortitude displayed by these heroic soldiers, when every one realized the cause was lost and the end must soon come, I quote from General Lee's report of this Hatcher Run fight, dated 8 February, 1865: "Yesterday, the most inclement day of the winter, the troops had to be retained in line of battle, having been in the same condition the two previous days and nights. I regret to be obliged to state that under these circumstances, heightened by the assault and the fire of the enemy, some of the men were suffering from reduced rations and scant clothing, exposed to battle, cold, hail and sleet. I have directed Colonel Cole, chief commissary, who reports that he has not a pound of meal at his disposal, to visit Richmond and see if something cannot be done. If some change is not made, and the Commissary Department not reorganized, I apprehend dire results. The physical strength of the men, their courage, services, must fail under this treatment. Our cavalry has to be dispersed for the want of forage. I had to bring Wm. H. F. Lee's Division forty miles Sunday night to get him in position." President Davis endorses this report as follows: "This is too sad to be patiently considered, and cannot have occurred without criminal neglect or gross incapacity. Let supplies be had by purchase or borrowing, or other possible mode."

APPOMATTOX, 9 APRIL, 1865.

On 28 March, 1863, General Fitzhugh Lee was ordered to move his division of cavalry, then on the extreme left of the Confederate lines in front of Richmond on the north side of the James river, to Sutherland's Station on the south side of the railroad, 19 miles from Petersburg, which he reached on the 29th, and next day marched towards Dinwiddie Court House, via Five Forks.

On 29 March, 1865, General Lee advises Secretary of War, General John C. Breckenridge, that "the enemy have crossed Hatcher's Run with a large force of cavalry and infantry and artillery."

On 1 April "that General Pickett, with three of his own and two of General Johnson's (Bushrod) Brigades, supported the cavalry under General Fitz. Lee, at Five Forks; that General Pickett forced his way to within less than a mile of Dinwiddie Court House, but later a large force, believed to be the Fifth Corps (Warren's), with other troops, turned Pickett's left and drove him back on the White Oak Road and separated him from General Fitz. Lee, who was compelled to fall back across Hatcher's Run; General Pickett's present position not known."

On 1 April, Longstreet was ordered with two of his divisions to the south side, and General W. N. Pendleton, chief of Artillery, was ordered at 8 p. m. to withdraw all his guns, which he in his report says, "was accomplished with great success, only sixty-one guns and thirteen caisons of the 250 field pieces belonging to the army on the lines near Richmond and Petersburg remained behind."

On 2 April (received at 10:40 a. m.) General Lee dispatches President Davis: "I see no prospect of doing more than holding our position here till night." Later on same day (received at 7 p. m.): "It is absolutely necessary that we should abandon our position tonight, or run the risk of being cut off in the morning."

General R. S. Ewell in his report, says: "At 10 a. m. Sunday (2 April, 1865), received message to return to the city of Richmond, and on doing so received the order for the evacuation and to destroy the stores that could not be moved. A mob of both sexes and all colors soon collected, and about 3 a. m. (3 April) they set fire to some buildings on Cary street, and began to plunder the city. I then ordered all my staff and couriers to scour the streets and sent word to General Kershaw, whose command was garrisoning Fort Gilmer, on the lines north of Richmond, to hurry his leading regiment into town. By daylight the riot was subdued, but many buildings which I had carefully directed should be spared, had been fired by the mob. By 7 a. m. the last troops had reached the south side, and Mayo's and the railroad bridges were on fire. I am convinced the burning of Richmond was the work of incendiaries."

On the afternoon of 6 April, Lieutenant-General Ewell and Major-General G. W. C. Lee, and their commands, were captured.

On the night of 7 April General Grant sent a note to General Lee, asking his surrender, to which General Lee replied, the time for surrender had not come. General Lee was still in hopes he could reach Appomattox Court House and there obtain supplies, and thence push on behind the Staunton river, and eventually unite with General Joseph E. Johnston somewhere in North Carolina. General Lee, with the remnant of his army, reached the neighborhood of Appomattox Court House on the evening of 8 April, but Sheridan's cavalry had gotten there first, captured the trains with the supplies, and obstructed Lee's further advance.

On the morning of the 9th, General Lee sent a flag of truce to General Grant, asking for an interview, and the same morning the two Generals met in the house of Mr. Wilmer McLean, in the village of Appomattox Court House, and the terms of the surrender were agreed upon. These were that the men and officers were to be paroled on a pledge not to take up arms again until properly exchanged. The officers were to retain their side arms, private horses and baggage. Those enlisted men who owned the artillery and cavalry horses or mules they were using, were also allowed to retain them. General Grant saying he supposed "most of the men in the ranks were small farmers who would need their horses to put in a crop to carry themselves and families through the next winter." It required several days to parole those surrendered, (some escaped to join Johnston's army and refused to surrender) and then, in groups and squads, or one by one, the paroled men dispersed to reach their homes as best they could. Thousands of them were penniless Many had hundreds of miles to travel, without money or means of transportation, but there was no rioting or outrage as they moved through the land, everywhere desolated and despoiled, to find their homes, in many cases, laid waste and destroyed. The same constancy and devotion to their country which had sustained them amid battle and strife unparalled, nerved them to face courageously this dark time of defeat and disappointment and to do their best to retrieve the widespread ruin of their beloved South."

In these last days of the war, the Twenty-sixth Regiment sustained severe losses in killed and wounded. Lieutenant J. W. Richardson was killed at Reams Station, and at Five Forks (1 April, 1865) Captain Thomas Lilly, who had succeeded Captain J. C. McLauchlin as Captain of Company K, and been put in command of the brigade sharpshooters, was killed. He was one of the best officers in the regiment. Colonel Lane, during the winter of 1864-5, suffered much from his wounds, especially the one in the neck and face, and about the middle of March went to the hospital at Salisbury for treatment. He was there when General Lee surrendered, and on 2 May, 1865, was paroled at Greensboro, N. C., with Johnston's army.

Lieutenant-Colonel Adams took command of the regiment after Colonel Lane went to the hospital, and except a few days on the retreat when he was temporarily in command of the brigade, was with his regiment. In his absence Captain T. J. Cureton, of Company B, commanded the Twenty-sixth, and surrendered the regiment at Appomattox. Lieutenant-Colonel Adams, however, signing the paroles.

NUMBERS PAROLED AT APPOMATTOX.

On 1 March, 1865, the Brigade Inspector reported the strength of MacRae's Brigade, present and effective for the field:

Officers	55
Enlisted men	1,119
Total	1,174

The capitulation rolls at Appomattox showed:

Heth's Division.	Officers.	Enlisted Men.
Major-General Harry Heth and Staff	15	...
John R. Cooke's Brigade	70	490
Joseph R. Davis' Brigade	21	54
Wm. MacRae's Brigade	42	400
Wm. McConnel's (formerly Archer's and Thomas')	54	426

The rolls for the entire army surrendered by General Lee:

	Officers.	Enlisted Men.
General Headquarters	69	212
Infantry	2,235	20,114
Cavalry	134	1,425
Artillery	184	2,392
Miscellaneous	159	1,307
Total	2,781	25,450–28,231

The number surrendered by the several regiments of MacRae's Brigade:

Eleventh Regiment, commanded by Colonel Wm. J. Martin, 74 muskets.

Twenty-sixth Regiment, commanded by Lieutenant-Colonel J. T. Adams, 120 muskets.

Forty-fourth Regiment, commanded by Major C. M. Stedman, 74 muskets.

Forty-seventh Regiment, commanded by Captain J. H. Thorpe, 72 muskets.

Fifty-second Regiment, commanded by Lieutenant-Colonel E. Erson, 60 muskets.

There was but one regiment in Heth's division that surrendered more muskets than did the Twenty-sixth, and that was the Fifteenth North Carolina Regiment, in Cooke's Brigade, which surrendered 122 muskets. In Major Moore's "Roster of North Carolina Troops" the aggregate of numbers enrolled in the Twenty-sixth Regiment is put down as 1,898, which is more than was enrolled in any regiment furnished the Confederate armies from North Carolina, according to said Roster.

RECAPITULATION OF THE COMMISSIONED OFFICERS OF THE REGIMENT.

(The field officers and captains are mentioned in the order of the date of their commissions; but the Lieutenants alphabetically, and their relative rank is not set out, as it is impossible in all cases to give.)

Colonels—Z. B. Vance, H. K. Burgwyn, Jr., John R. Lane.

Lieutenant Colonels—H. K. Burgwyn, Jr., John R. Lane, John T. Jones, James T. Adams.

Majors—Abner B. Carmichael, N. P. Rankin, James S. Kendall, John T. Jones, James T. Adams.

Adjutants—James B. Jordan. Acting at different times as Adjutant, Lieutenants John A. Polk, A. R. Johnson, Wm. N. Snelling.

Surgeons—Thomas J. Boykin, Llewellyn P. Warren.

Assistant Surgeons—Daniel M. Shaw, Wm. W. Gaither. Acting at different times as Assistant Surgeon, Captain W. S. McLean, Lieutenant George C. Underwood.

Regimental Quartermaster—Captain Joseph J. Young.

REGIMENTAL COMMISSARY—Captain Robert W. Goldston, Phineas Horton.

SERGEANT MAJORS—L. L. Polk, Montford S. McRae, John E. Moore.

QUARTERMASTER SERGEANT—Abram J. Lane.

COMMISSARY SERGEANT—Jesse F. Ferguson.

ORDNANCE SERGEANT—E. H. Hornaday.

HOSPITAL STEWARD—Benjamin Hind.

CHAPLAINS—Rev. Robert H. Marsh, Richard Nye Price, Styring S. Moore, John Huske Tillinghast.

COMPANY A—Captains, A. N. McMillan, Samuel P. Wagg, A. B. Duvall. Lieutenants, A. B. Duvall, J. M. Duvall, L. C. Gentry, J. B. Houck, James Porter, George R. Reeves, Jesse A. Reeves.

COMPANY B—Captains, J. J. C. Steele, William Wilson, Thomas J. Cureton. Lieutenants, A. Brietz, Taylor G. Cureton, Thos. J. Cureton, Calvin Dickinson, Wm. M. Estridge, John W. Richardson, Wm. W. Richardson, Wm. Wilson.

COMPANY C—Captains, A. B. Carmichael, A. H. Horton, Thos. L. Ferguson, J. A. Jarrett. Lieutenants, Wm. W. Hampton, John M. Harris, A. H. Horton, Rufus D. Horton, Phineas Horton, J. A. Jarratt, Wm. Porter.

COMPANY D—Captains, Oscar R. Rand, James T. Adams, Gaston H. Broughton. Lieutenants, James T. Adams, Gaston H. Broughton, James G. M. Jones, James B. Jordan, Wm. Snelling, James W. Vinson, M. J. Woodall.

COMPANY E—Captains, W. S. Webster, Stephen W. Brewer. Lieutenants, Stephen W. Brewer, Bryant C. Dunlap, John R. Emerson, Orran A. Hanner, Wm. J. Headen, W. J. Lambert, E. H. McManus.

COMPANY F—Captains, N. P. Rankin, Joseph R. Ballew, Romulus M. Tuttle. Lieutenants, Joseph R. Ballew, Abner B. Hayes, John B. Holloway, R. N. Hudspeth, Alfred T. Stuart, Charles M. Sudderth, R. M. Tuttle.

COMPANY G—Captains, W. S. McLean, John R. Lane, H. C. Albright, A. R. Johnson. Lieutenants, H. C. Albright,

A. R. Johnson, Wm. G. Lane, J. A. Lowe, John E. Matthews, Samuel E. Teague, George C. Underwood.

Company H—Captains, Wm. P. Martin, Clement Dowd, J. D. McIver, M. McLeod. Lieutenants, Clement Dowd, Robert W. Goldston, J. H. McGilvery, James D. McIver, M. McLeod, George Willcox.

Company I—Captains, Wilson A. White, John T. Jones, N. G. Bradford. Lieutenants, M. B. Blair, N. G. Bradford, John Carson, Rufus Deal, S. P. Dula, J. C. Greer, John T. Jones, J. G. Sudderth.

Company K—Captains, James C. Carraway, John C. McLauchlin, Thomas Lilly. Lieutenants, Wm. C. Boggan, J. L. Henry, Wm. L. Ingram, James S. Kendall, Thomas Lilly, John C. McLauchlin, J. A. Polk.

The casualties in the regiment among the above officers from first to last were as follows:

KILLED.

Colonel H. K. Burgwyn, Jr., Lieutenant-Colonel John T. Jones, Major Abner C. Carmichael, Captains Albright, Lilly, Martin, Wilson and Wagg; Lieutenants J. M. Duvall, Deal, Emerson, Hayes, Henry, Holloway, John W. Richardson, William W. Richardson, C. M. Sudderth, Teague, Woodall—19.

WOUNDED.

Colonel John R. Lane, Lieutenant-Colonel James T. Adams, Adjutant James B. Jordan; Captains Bradford, Brewer, Broughton, Cureton, A. B. Duvall, Jarrett, McLauchlin, McLeod, McMillan, Tuttle; Lieutenants Brietz, Estridge, Gentry, Green, Hanner, R. D. Horton, Houck, Hudspeth, Ingram, J. G. M. Jones, Lambert, W. G. Lane, Lowe, McGilvery, McManus, Polk, Porter, Snelling, Willcox—32.

Many of the above were wounded more than once.

CHIEF SAMUEL T. MICKEY'S BAND.

A history of the Twenty-sixth Regiment would not be complete without an account of its band, regarded as one of the best in the Army of Northern Virginia. It was recruited chiefly from Salem, N. C., and most of its members belonged to a band in that town prior to the war. Samuel Timothy Mickey, of Salem, was the leader, and the names of the other members are as follows: A. P. Gibson, J. A. Lineback, H. A. Siddell, W. H. Hall, Julius A. Transon, Charles Transon, A. L. Hauser, A. Meinung, W. A. Lemly, D. T. Crouse, J. O. Hall, W. A. Reich, D. J. Hackney, Edward Peterson. Only one of them died during the war, viz., A. L. Hauser.

Captain Mickey still leads a band in Salem, and is a prosperous mechanic. W. A. Lemley is the president of the Wachovia National Bank, of Winston, N. C., and J. D. Hackney is a Baptist Preacher.

The band was recruited for Wheeler's Battalion, but at the capture of that command at Roanoke Island, Captain Mickey went to New Bern to seek employment. He thus describes his first meeting with Colonel Vance: "I was sitting in the lobby of the Gaston House, New Bern, when a man wearing a Colonel's uniform came in with a loaf of bread under each arm. This was Zeb Vance. I spoke to him and told him my errand. Colonel Vance replied: 'You are the very man I am looking for. You represent the Salem band. Come to my regiment at Wood's brick yard, four miles below New Bern.' Next morning (March, 1862), I went down to the camp, was met by Captain Horton, of Company C, and as the result of my visit, the band was engaged and at first it was paid by the officers." The members being musicians of unusual cultivation and intelligence, under Captain Mickey's indefatigable labors, the band soon acquired great celebrity and was in constant demand for serenades and military parades. On the Sunday before Gettysburg, at Fayetteville, Pa., Chaplain Wells preached before the Brigade. His text was "The Harvest is past, and the Summer is ended and we are not saved." It was an eloquent discourse and made a great impression. After the services were over, and the band returned to its quarters, the drummer (W. A. Reich) remarked: "Boys, I believe we are going to lose our Colonel in the next fight. Did you notice his looks during the sermon?" Captain Mickey replied: "Yes, I did; he looked right serious." As appears above in this history their Colonel was lost to them in the next fight.

Captain Mickey thus writes of Gettysburg: "The Yankees were in three lines on the hill pouring volley after volley on our men as they came through the fields. The color guard were all shot down, the colors fell fourteen times. Colonel Burgwyn was shot down with the colors, and Captain McCreery of General Pettigrew's staff, was also killed with the colors. General Pettigrew said the men of the Twenty-sixth shot as if they were shooting at squirrels; that their shots counted. After the first day's battle, Colonel Marshall, commanding the brigade, sent an order for the bands of the Twenty-sixth and Eleventh Regiments to report to his headquarters, that the men were anxious to hear some music. The two bands played numerous pieces which seemed to enliven and cheer the soldiers. While the bands were playing, they were shelled by the enemy, and as they left a shell burst just where they had been standing.

"On the retreat from Gettysburg to Bunker Hill, the band serenaded General Lee and other officers. After the serenade to General Lee, Colonel Taylor, his Adjutant General, came out of his tent and made a little talk. Thanked the band for the serenade, and said he didn't know how they would get along without bands; that they cheered up the men so much; that he noticed the style of our music was different from that of the other bands in the army." Mr. W. H. Hall was captured near Green Castle on this retreat.

Just before the campaign of the Wilderness opened, Colonel Lane took his band in a four-horse wagon to serenade General Lee at night. The Colonel was invited into General Lee's tent while the music was playing. General Lee remarked that we would not be idle many days; that Grant was making preparations to cross; and General Lee then said if he could only strike him with his center, he though he would be able to make him recross in a way not so pleasant as was his coming over. "I can re-enforce from each wing," said General Lee.

Later on in the conversation, General Lee remarked: "I don't believe we can have an army without music."

During the Spring of 1862-'63, and the winter of 1863-'64, the band was granted a furlough and gave several concerts in different parts of the State, and everywhere met with the most enthusiastic reception. They played at Governor Vance's first inauguration.

The band remained with the regiment to the end and was captured on the retreat from Petersburg and taken to City Point, and thence to Point Lookout. They were finally released, and Captain Mickey reached home (Salem) on 3 July, 1865.

DESERTIONS.

A few words on this subject is of historical interest. Except in the closing days of the struggle, there were few, if any, desertions to the enemy. There were numerous cases of absence without leave, but the parties did not mean to desert their colors. Impelled by an irresistible yearning to see those they had left behind in their humble homes, they would go home without leave, but when this longing was gratified, they would voluntarily rejoin their commands and do as loyal service as any.

It became finally necessary to visit the death penalty in instances, as an example to deter others. Sergeant Andrew Wyatt, Company B, and some ten others of the regiment on 10 December, 1862, deserted while the regiment was stationed at Garysburg, N. C. They started for their homes in the Western part of the State, but were arrested at a crossing on the Roanoke river. The Sergeant was court-martialed, convicted and condemned to be shot. While in camp near Magnolia, N. C., January 1863, he was taken out in a wagon to the place of his execution, where the brigade was drawn up in a three-quarter square to witness the shooting. The prisoner was blindfolded, ordered to kneel down by the freshly dug grave, the firing squad stood with their guns at a "ready" and the officer was reading the sentence, when an orderly rode up with an order from General French, commanding the department, granting a pardon. Subsequent to his conviction the officers of the regiment became satisfied that the Sergeant only intended to go home and see his family, and then return to his command, and on their request, his life was spared. Sergeant Wyatt was killed at Gettysburg, bravely doing his duty in that famous first day's battle.

While at Hanover Junction in June, 1863, Colonel Lane was president of a general court-martial. Several of his regiment had been tried for desertion and sentenced to be shot, and were awaiting their execution. Among them was John Vinson, a member of Colonel Lane's old company (Company G). When the regiment started for Pennsylvania these prisoners were marched at the rear of the regiment under guard. Riding by their side one day, Colonel Lane remarked to them: "Are you in sympathy with the South, and if permitted to do so, will you help us fight in this next battle?" They said: "We will. We only wished to go home to see our folks." General Lee informed of this, ordered them restored to duty, and no soldiers fought better at Gettysburg. John Vinson was wounded with the colors of the regiment, having volunteered to carry them. S. T. Dula, of Company I, was recommended by Major Jones for promotion for gallant conduct at Gettysburg, where he was wounded.

After the return to Virginia, he deserted, but voluntarily returned to the regiment after an absence of two or three weeks. Major Jones sent for him and said to him: "What in the world did you mean by doing this. You have put me in a devil of a fix." Dula replied that "he heard his wife had had a little one, and he could not resist going home to see it." He was allowed to go on duty, and was killed at Bristoe Station, leading the charge.

Governor Vance was most energetic in getting these "absent without leave" men to return to their commands. He issued several proclamations on the subject. In the proclamation dated 27 January, 1863, he promised to use his influence with the authorities to pardon all those who would return to duty voluntarily. Many returned to their commands in response to this proclamation, and General Lee writes Governor Vance under date of 26 March, 1863: "I at once remitted the penalties inflicted by the courts, and restored the men to duty. I also directed that no charges should be preferred against soldiers who returned to duty under the promises contained in your proclamation."

Governor Vance ordered the militia officials to assist the Confederate authorities in arresting those who continued to remain away without leave. On one occasion there was a fight between his militia officers and some deserters resisting arrest, in which one of the militia was killed. The deserter who did the killing was arrested and a habeas corpus was sued out before Chief Justice Pearson, of the State Supreme Court, who discharged the prisoner on the ground that the militia had no authority to arrest a deserter from the Confederate army. This first proclamation was followed by two others dated 11 May and 24 August, 1864. In this last one, Governor Vance gives this notice: "Warning is hereby given that in all cases where either Civil Magistrate or Militia, or home guard officers refuse or neglect faithfully to perform their duties in the arrest of deserters, upon proper evidence submitted to me, the Executive protection extended to them under Acts of Congress (Confederate) shall be withdrawn, as I cannot certify that officers, Civil or Military, who refuse to perform their duties are necessary to the administration of laws which they will not execute."

MORALE OF THE CONFEDERATE SOLDIER.

In his Personal Memoirs, General Grant, writing of the conduct of the Confederate troops as late as 6 April, 1865, three days before the surrender at Appomattox, uses these words: "There was as much gallantry displayed by some of the Confederates in these little engagements as was displayed at any time during the war, notwithstanding the sad defeats of the past week." On that day (6 April, 1865), Colonel Washburn with two regiments of infantry and eight of cavalry, under Colonel Read, of General Ord's Staff, with orders to destroy the High Bridge over the Appomattox river near Farmville, returning from the expedition, met the advance of a detachment of the Confederate army on its retreat marching in the same direction. Colonel Washburn gave the order to charge. It was unsuccessful. Colonels Washburn and Read were mortally wounded, nearly every officer and most of the rank and file were killed or wounded, and the balance were captured.

Finally as his reasons for surrender, General Lee says: "On the morning of 9 April, 1865, there were 7,892 organized infantry with arms, 63 pieces of artillery, and 2,100 cavalry. We had no subsistence for man or horse, and it could not be gathered in the country, and the men deprived of food and sleep for many days were worn out and exhausted."

A member of the regiment thus writes under date of 3 August, 1900: "The morale, the elan, the physique of the Twenty-sixth, has not been equalled. My greatest glory is that I was so intimately associated with its history."

We will bring this history to a close by a short biographical sketch of some of those through whose labors and military skill the regiment was brought to that state of high efficiency which enabled it to accomplish such feats of arms as will for all time set it apart as one of the most famous military commands in the annals of war.

The youthfulness of the officers of the regiment was remarkable. Colonel Burgwyn's class at the Virginia Military Institute was not to have graduated until June, 1861, but was graduated in April previous, to enable its members to offer their services in the war then inevitable between the United States and the New Confederacy of Southern States, organized at Montgomery, Ala., February, 1861.

Lieutenant-Colonel John T. Jones was to have graduated at Chapel Hill (University of the State) in June, 1861, but volunteered in a company organized at Chapel Hill in the Spring of 1861, that became Company D, of the Bethel Regiment.

Captains Wilson, Albright, Tuttle, and McLaughlin, also left college prior to their graduation, to join the army.

Colonel Vance was thirty-one years old and Colonel Lane twenty-six when they volunteered. Lieutenant-Colonel Adams had barely attained his majority when he was elected Third Lieutenant in the Wake Guards, and Captains Wilson, Lilly, Broughton, Cureton, Duvall, and the company officers, almost without exception, were under twenty-five years of age when they volunteered.

COLONEL ZEBULON BAIRD VANCE.

The civic career of this distinguished citizen of North Carolina appears in so many publications we will confine our remarks entirely to his military record. A member of the regiment thus speaks of him as a soldier: "I remember well the first time I ever saw him. He had no appearance in the world of a soldier; his hair was long and flowing over his shoulders, and he was wearing a little seal skin coat, from which I judged him to be a Chaplain. He had not long been absent from the hustings of Western North Carolina, and had but little experience in war as Captain in the Fourteenth Regiment. When he came to the camp he soon began to display the same qualities which made him so popular all over our State.

"In the first place he had the keenest sympathy with his men. They soon came to feel that Colonel Vance loved them, and made their troubles his own. In the next place, Colonel Vance was able to inspire his men with the belief that he had confidence in them. These two essentials to a good commanding officer were, perhaps never possessed by any man to a greater degree than by Colonel Vance.

"In drill and discipline, Colonel Vance was at first deficient. I mention this not in any way to discredit him, for his life as a politician had given him no opportunity to develop these essentials in the character of an officer.

"I mention the fact to show the wisdom he displayed in the matter, for when he saw his regiment deteriorating, he recognized his deficiency and set about to correct it. He turned to his Lieutenant-Colonel, Harry King Burgwyn, who had been trained at the Virginia Military Institute, and was a very master of drill and discipline. He put himself and his subordinates under the tutorship of this brilliant young officer. The result was most satisfactory. Colonel Vance and many of his officers soon became well schooled in the methods of drill and discipline, and his regiment became almost a perfect instrument of war, devoted to their commander. In battle I always marked him as cool and coura-

J. R. Lane. H. K. Burgwyn. Z. B. Vance.
Three Colonels of the 26th N. C. Regiment.

geous. When duty called Vance from the army to be Governor of North Carolina, in the most trying period of the war, he had gained much from his career as Colonel of the Twenty-sixth that I believe he found valuable in his future duties. He had a sympahtetic knowledge of the needs of the Confederate soldier, the war wrought into his sinews; he knew how with all his kindness to deal firmly with men and affairs. He was a better Governor for having been Colonel."

COLONEL HARRY KING BURGWYN, JR.

A short time after the death of this young officer, born 3 October, 1841, probably the youngest of his rank in the Confederate army—obituary notices appeared in the Raleigh papers. From one of them we copy: "It would be unjust to the living no less than to the memory of the young hero and martyr who now sleeps beneath the sod of a distant and foreign State, were the death of Colonel Harry King Burgwyn, Jr., permitted to pass with the brief notice of his fall published in a late number of this paper.

"The life, career and death of young Burgwyn, convey a lesson to the youth of this Confederacy which cannot be too well studied and thoroughly profited by. He was the eldest son of Henry King Burgwyn, Esq., of Northampton County, in this State, his mother was Miss Anna Greenough, of Boston, Mass., and had barely attained the age of twenty-one years when he attested his love for his country by the sacrifice of his life on the altar of its liberties. Born to the enjoyment of affluence, he might, as too many of our youth do, have been content to grow up in idleness and luxurious ease. But such a life had no charms for him. Blessed with a fine capacity and docile disposition, he well availed himself of the abundant means of education afforded him by his parents.

"His education preparatory to his entrance into the University of the State, was partly from private tutors in the family and at Burlington, New Jersey, and at West Point, where he was a private pupil of Foster,—now the Yankee General at New Bern. Leaving West Point, he entered the University of his State, and graduated with the highest honors (1859). At this period he might, as the phrase goes, have been considered 'educated.' Not so, however, thought his father. Foreseeing the difficulties which have culminated in a war between the South and the North, and desirous that his son should be prepared for usefulness in every emergency, he placed him in the Virginia Military Institute, where he was when hostilities commenced. Of the course of young Burgwyn in that institution an idea may be formed from the following letter from the now lamented Stonewall Jackson:

"LEXINGTON, VA., April 16, 1861.

SIR:—The object of this letter is to recommend Cadet H. K. Burgwyn, of North Carolina, for a commission in the artillery of the Southern Confederacy. Mr. B. is not only a high-toned Southern gentleman, but in consequence of the highly practical as well as scientific character of his mind, he possesses qualities well calculated to make him an ornament not only to the artillery, but to any branch of the military service.

T. J. JACKSON,
Prof. Nat. Phil. and Instr. Va. M. I.

To L. P. Walker, Secretary of War.'

"The discriminating and sagacious judgment of the professor has been fully attested by the career of the pupil from the moment he entered the service to the day on which he met a soldier's fate on the bloodiest field of the war, as with colors in hand, he was leading his men on to victory. When New Bern fell, he was the last man of his regiment to cross the creek on the retreat—having refused to enter the boat until all were safely passed over. On this occasion young Burgwyn was Lieutenant-Colonel of the regiment, the Colonel being the present Governor, Vance.

"From this State we follow the subject of our narrative to the bloody fields around Richmond, winding up with the terrific fight at Malvern Hill, in which his regiment was unsurpassed for heroism by any troops on the field.

"On the resignation of Colonel Vance, when he became Governor-elect of the State, young Burgwyn was promoted Colonel, and soon thereafter we find him again in service in his native State. In the critical campaign in Martin County, when the enemy were threatening disastrous consequences to the region of the Roanoke river, we find Colonel Burgwyn performing signal services, especially in the engagement of Rawls' Mills, where he displayed a cool judgment and indomitable courage of which a veteran of many years standing might have been proud. In all the course of this career, so well calculated "to turn the head" of one so young, Colonel Burgwyn displayed a modesty so commendable that he silenced the tongue of envy and won the confidence of his brothers in arms. When on Governor Vance's resignation, it was suggested that he was too young for the Colonelcy, General D. H. Hill wrote of him: 'Lieutenant-Colonel Burgwyn has shown the highest qualities of a soldier and officer, in camp and on the battle field, and ought by all means to be promoted.'

"As we have seen, Colonel Burgwyn did receive the promotion and subsequently was strongly recommended for the higher command of Brigadier-General.

"We have thus given a brief sketch of the career of one whose death in the very outset of manhood prompts the question, 'If he was such in the gristle, what would he not have been in the bone?' "

His last words after sending a farewell to his parents and family were: "Tell the General my men never failed me at a single point." *"Felix non solum claritate vitae, sed etiam opportunitate mortis."*

In a letter from Major George P. Collins, Brigade Quartermaster, written from the battle field and dated 3 July, 1863, and addressed to Colonel Burgwyn's father at Raleigh, N. C., he thus describes the end: "Captain J. J. Young (Regimental Quartermaster) has undertaken to give you the sad news of your son's death, but I cannot let the opportunity pass without expressing my deep sympathy with his bereaved parents and family, as well as testifying to the gallant and soldierly manner in which he met his death. He was one of eleven (afterwards ascertained to be fourteen) shot bearing the colors of his regiment, and fell with his sword in his hand, cheering his men on to victory. The ball passed through the lower part of both lungs and he lived about two hours. Among his last words he asked how his men fought, and said they would never disgrace him. He died in the arms of Lieutenant Louis G. Young (Aide de Camp to General Pettigrew) bidding all farewell and sending love to his mother, father, sister and brothers." He was buried under a walnut tree (a gun case answering for a coffin) by Major Collins and Captain J. J. Young, assisted by M. F. Boyle, of Company B, the regimental mail carrier, and by Jesse T. Ferguson, of Company C, the regimental Commissary Sergeant. In the Spring of 1867 his remains were brought from Gettysburg, and re-interred in the Soldier's Cemetery at Raleigh, where he rests in the midst of his comrades who wore the gray, and who, like him, gave up their lives in the defense of a cause they believed holy and just. A handsome monument erected by his parents marks the grave.

On 20 October, 1897, a portrait of the "Three Colonels of the Twenty-sixth Regiment," on one canvass, was presented to the State with appropriate ceremonies. The presentation took place during Fair Week, and was held in the Central Hall of the main building at the Fair grounds.

COLONEL JOHN RANDOLPH LANE.

This battle scarred veteran still lives (April, 1901) in vigorous manhood. He was born in Chatham County, 4 July, 1835, and is a direct descendant from Colonel Joel Lane, of Wake County, from whom the land on which the City of Raleigh is located was bought. General Joe Lane, the Vice-Presidential candidate in 1860 on the Breckinridge and Lane ticket, was his near relative.

He enlisted as a private in Company G and soon became Corporal. On the resignation of his Captain in the Fall of 1861, he was elected over the heads of all his commissioned officers, to command the company. He was re-elected Captain at the reorganization of the regiment in the Spring of 1862. At the battle of New Bern, Captain Lane was complimented for bravery and coolness under fire, and in the night attack on 25 June, 1862, upon his regiment while on picket, referred to in the body of this history, his company was one of the three which stood firm under such a trying ordeal.

On the promotion of Lieutenant-Colonel Burgwyn, to the

Colonelcy, the position of Major also being vacant, owing to the death of Major Kendall, Captain Lane was promoted over several senior captains to be Lieutenant-Colonel. After Gettysburg, he was made full Colonel, his commission bore date of 1 July, 1863, in recognition of his heroic conduct on that battle field. Seeing his Colonel fall, he immediately assumed command, and realizing that if the death of their Colonel was known it would have a depressing effect upon the men, he did not impart it to the regiment, but inspired his men with the cheering words that fell from the lips of his stricken commander, and seizing his flag, calls upon his men to follow him. All depended now on Colonel Lane. There is a line of the enemy yet to be broken, and there is only a handful of his men left to do the work. We have seen how the crisis was met and the glorious victory and its cost. General Pettigrew anxiously watching the contest, when he saw the enemy giving way on their last line before this desperate charge of the regiment, with Colonel Lane at the head, exclaims: "It is the bravest act I ever saw." As described in the body of this article, Colonel Lane was thought to be mortally wounded, but escaping capture, he returned to duty in the Fall of 1863. Wounded at the battle of the Wilderness, 5 May, 1864, he refused a furlough. Again wounded in right leg at Yellow Tavern, south of Petersburg, in summer of 1864, but refused to leave the field. At Reams' Station 25 August, 1864, he was wounded in left breast, just over the heart by a piece of shell, fracturing two ribs and breaking one, and tearing open the flesh to the bone. Supposed to be mortally wounded, he wonderfully recovered and returned to duty November, 1864; remained in command until broken down by exposure and suffering from his wounds, he went to the hospital for treatment, and was at Danville, Va., when the remnant of his heroic regiment surrendered at Appomattox. He was paroled at Greensboro, N. C., on 2 May, 1865, and returned to his home to take up the struggle for a living he had laid aside four years before.

Since the war Colonel Lane has become a prosperous merchant and large land owner in his native county, all accumulated by his untiring energy, business ability and thrift. He is conspicuous for his liberality and devotion to the old comrades of his immortal regiment.

LIEUTENANT-COLONEL JOHN THOMAS JONES.

Was born in Caldwell County, N. C., on 21 January, 1841. In 1857 he entered the University of North Carolina and there remained until the breaking out of the war between the States. During his senior year, and just prior to his graduation, he volunteered as a private in the Orange Light Infantry commanded by Captain R. J. Ashe, which company became Company D in the "Bethel" Regiment. He was with his regiment at the battle of Big Bethel, and after its term of service expired, came home to Caldwell County and engaged actively in enlisting that body of men which became known to fame as Company I, of the Twenty-sixth North Carolina Regiment of Infantry. Was elected Second Lieutenant, and upon the reorganization of the regiment for the war, was elected Captain; was promoted to be Major of the regiment when the noble Harry Burgwyn became Colonel, and after Colonel Burgwyn's glorious death, became Lieutenant-Colonel in place of Colonel Lane, who succeeded the gallant Burgwyn.

He passed through all the battles and combats in which his regiment was engaged, distinguishing himself especially at Rawls' Mills and Gettysburg. In the latter battle he received a wound, but he declined to leave the field, and commanded the regiment after the fall of Colonels Burgwyn and Lane, and was in command of the brigade at the close of the charge on the third day. At the great battle of the Wilderness, 6 May, 1864, after the wounding of Colonel Lane, he assumed command and was mortally wounded leading his regiment in a charge against overwhelming numbers. When told by Assistant Surgeon W. W. Gaither that his wound was mortal, says the Surgeon: "With a most yearning expression he replied, 'It must not be. I was born to accomplish more good than I have done.' "

After the battle of Gettysburg, where his younger brother, Walter, a private in Company I, was killed, Lieutenant-Colonel Jones, then Major, was for some time in command of the brigade, all the other field officers present at the battle having been killed or wounded. His remains, with those of his brother, rest in one grave in the family cemetery in the beautiful "Happy Valley" in Caldwell County. The John T. Jones Camp, U. C. V., of Lenoir, N. C., is named in honor of this brave soldier and meritorious officer. The friendship between Colonel Jones and Colonel Burgwyn was so marked that subsequent to their deaths one of the officers of the regiment composed some beautiful lines on "Colonels Harry, and John," likening them to Jonathan and David.

DESERVING OF SPECIAL MENTION.

Lieutenant-Colonel James T. Adams. This meritorious officer rose from Second Lieutenant in Company D, from Wake County, to be Lieutenant-Colonel of the regiment, and during the last days of the war was in command of the regiment and on the retreat from Petersburg, was at times in command of the brigade.

He was wounded through the hip at Malvern Hill and seriously through the shoulder at Gettysburg, and except while on furlough from wounds was never excused from duty. At Spottsylvania Court House, the brigade was ordered to drive the enemy from their position which menaced General Lee's rear and communications with Richmond. "The enemy had made a breastwork out of a fence in a piney old field and chinked the cracks between the rails with dry pine straw. As the brigade neared them, the enemy set fire to the fence and old field which burnt rapidly. Nothing daunted, the Confederates charged through the flames and over the burning fence, and drove their opponents in discomfiture from the field."

At Hancock's defeat at Burgess' Mill, on the Boydton plank road south of Petersburg, 27 October, 1864, Lieutenant-Colonel Adams in command of the regiment, acted with such conspicuous gallantry as to call forth the warm commendation of his brigade commander, General William MacRae. The brigade with other troops were ordered to dislodge Hancock, who had cut through the Confederate lines. The brigade charged the enemy in its front, drove him from his position, capturing a battery. The troops on our left failed to carry the lines in their front and the Federals closed in behind MacRae's Brigade and completely cut them off from their friends. The brigade reformed, about faced and charged, forcing their way through and in a hand to hand fight captured a battery and carried it out with them. In this action, the color-bearer of the Twenty-sixth Regiment was either shot down in the charge or got beyond eyesight in the dense swamp and undergrowth through which the men charged, and after it was over, the order was given to fall in on the colors of the Forty-fourth Regiment. Colonel Adams, who had lingered behind to see what had become of his color-bearer, ran out between the lines, and thinking his men a little downcast at losing their colors, he jumped up on a stump and called out, "Twenty-sixth, rally on your commander. He is here if his colors are lost." The men responded with a cheer.

At the brilliant victory of Reams' Station, after Colonel Lane was wounded, Lieutenant-Colonel Adams took command and was ever thereafter present with his regiment until its surrender at Appomattox, where he signed the paroles of his command.

Since the war Colonel Adams has resided in Wake County, a prosperous man in his business, respected and esteemed by all.

Dr. Thomas J. Boykin was Surgeon of the regiment, and remained with it until Colonel Vance's election as Governor, when he became Brigade Surgeon of Ransom's Brigade, and later was appointed State agent and sent to the Bermuda Islands, to handle blockade supplies for the State.

Dr. Boykin was born in Sampson County, N. C., in 1828, educated at Wake Forest College, and graduated at the Medical Department of the University of Pennsylvania. Practiced his profession in Kinston and Clinton, but removed to Nebraska Territory about the year 1856. Was elected a member of the upper branch of the Territorial Legislature. Immediately after the fall of Fort Sumter, (14 April, 1861) he returned to his native State and was appointed Surgeon of the Twenty-sixth Regiment.

Assistant Surgeon William W. Gaither. This officer who

most faithfully and acceptably served with the regiment until December, 1864, when he was promoted to be Surgeon of the Twenty-eighth North Carolina Regiment, graduated from the Jefferson Medical College of Philadelphia, in the class of 1860. Enlisted as a private in the "Hibriten Mountaineers," which became the afterwards famous Company F, in the Twenty-sixth Regiment. At first, serving as Hospital Steward, he was soon commissioned Assistant Surgeon, assigned to the regiment and put in charge of the hospital at Carolina City, below New Bern.

At Gettysburg, Dr. Gaither was all night getting the wounded from the field of the first day's fight and worked with them all the next day and night. On the afternoon of the third day, went to the regiment in line of battle. Under date of 5 September, 1900, Dr. Gaither writes: "I was on the field, saw the futile charge on the Cemetery wall, and the recoil. I got only three of the slightly wounded. When we got to Hagerstown, I went to sleep and slept for two entire days, so utterly exhausted I was."

Not one of the wounded who crossed the Potomac, but returned to duty sooner than any who before or after stopped in hospital. Fourteen patients marched all night in a big rain twelve miles, sick from three to twelve days with malarial fever, and none reported sick next morning. The doctor narrates this incident: "D. L. and R. C., members of Company I, from Caldwell County, had been fighting off and on during the day. About evening R. C. says to D. L., 'Demps, I'll hurt you directly,' and proceeded to knock him down and pulled out his right eye ball. D. L. did not even report sick. Two days after I found him lagging a little in the rear and asked him what was the matter. He said: 'R. C. had pulled his eye out, but it was all right now." While in camp at Garysburg, N. C., Fall of 1862, two patients with smallpox in third day of eruption, came to Surgeon's call wanting to know what caused the breaking out. They were not isolated and there were no new cases in the regiment, but more intense inflammation in all vaccinated arms.

In the winter of 1863-'64, while the army was in winter quarters around Orange Court House, Va., the number of men absent without leave at home became a matter of serious consideration, and the best way to put a stop to it was canvassed among the officers. There were several publications in the newspapers on the subject, and Assistant Surgeon Gaither wrote a set of resolutions which were passed by the officers in meeting, which attracted general notice and were universally approved as the best presentation of the situation that appeared.

Captain Joseph J. Young, A. Q. M. This gentleman had an unique experience as a soldier. He was the regimental quartermaster from the beginning to the close, and no command was ever blessed with a better one. He was wrapped up in his regiment and he could not do too much for them. He has kept copies of the regimental muster and pay rolls of the regiment which he treasures as among his most valuable possessions to be bequeathed to his children. In the latter months of the war when the number of the regimental quartermasters was reduced to two to a brigade, he and Captain John Gatlin, Fifty-second Regiment, were retained for MacRae's Brigade, and thus in addition to the care of his regiment, the brigade also received the benefit of Captain Young's valuable services and experience, and he always acted brigade quartermaster in the absence of Major Collins.

It was Captain Young's timely information, carried to Colonel Vance at the Captain's great personal risk, during the battle of New Bern, which advised Colonel Vance of the retreat of the other troops in time to enable the Colonel to withdraw the Twenty-sixth Regiment from the works and escape capture. We have seen how prompt Captain Young is to write his old Colonel the day after the battle of Gettysburg, of the glorious record this regiment made on that gory stained field; and as he began his military career with them, so at the end he was one of his immortal regiment to surrender at Appomattox.

Captain Young was born in Wake County, 1 January, 1832, and in May, 1861, he enlisted in Captain O. R. Rand's Company D, in the Twenty-sixth Regiment; was appointed by Colonel Vance Quartermaster of his regiment.

In December, 1864, Captain Young was sent to Eastern North Carolina to collect and forward supplies to Lee's army.

Adjutant James B. Jordan was born in Raleigh, N. C., 8 June, 1836. He was in business in Tennessee when on the secession of South Carolina, he returned to his native State and was elected First Lieutenant in Company D, of the Twenty-sixth Regiment and at the organization he was appointed Adjutant.

This position he held with honor and distinction until in the third day's fight at Gettysburg, he was seriously wounded, taken prisoner and carried to Johnson's Island, where he was detained as a prisoner until the close of the war.

In 1888, he was made Clerk of the Circuit Court of Volusia County, Florida, which position he held at his death, 27 April, 1899.

Captain Samuel P. Wagg, Company A. This gallant young officer was killed in the charge of Pettigrew on the third day at Gettysburg, within a few feet of the enemy's works. When the call for troops was issued at the breaking out of hostilities, he promptly enlisted in the first company that was organized in his county (Ashe) and was elected its First Sergeant. At the reorganization of the regiment in the Spring of 1862, he was elected Captain and was ever at his post of duty. Captain Wagg was buried on the field.

Captain Thomas J. Cureton, Company B. This officer succeeded to the command of Company B on the death of the gallant Captain William Wilson, killed on the first day's fight at Gettysburg.

Lieutenant Cureton was himself wounded on the third day in the shoulder, but declined to leave the field, and assisted in reforming the brigade as its shattered remnants recoiled from the assault on Cemetery Heights.

Captain Cureton was again wounded at Hanover Junction on 23 May, 1864, while in command of the skirmish line, but returned to duty in December, 1864, and remained with his regiment until the close, and much of the time was in command of it on the retreat to Appomattox, when Colonel Adams was in command of the brigade.

Before the war, Captain Cureton was a farmer, living in Union County, N. C. His grandfather owned the property in the Waxhaw settlement, North Carolina, where Andrew

Jackson was born, and where Captain Cureton's father was born. Since the war, Captain Cureton has resided in Charlotte, N. C., and Fort Mills, S. C., engaged in business as a cotton merchant, and now lives at Windsor, S. C.

Captain Stephen W. Brewer, Company E, was born in Chatham County 26 September, 1835; enlisted in Company E, Twenty-sixth North Carolina Regiment; was elected Third Lieutenant when the company was organized, and at its reorganization in the Spring of 1862, was elected Captain.

After the first day's fight at Gettysburg, in which his company lost 18 killed and mortally wounded, and 52 wounded, he led the twelve remaining into the third day's fight, that historic, but disastrous charge of Pickett and Pettigrew, and lost all but two killed and wounded. Captain Brewer was shot down, badly wounded, carrying his regiment's flag and fell near the enemy's line.

He was captured at Greencastle, Md., on the retreat from Gettysburg, and was confined as a prisoner of war in different Federal prisons, chiefly at Johnson's Island, Ohio, until March, 1865, when he was paroled.

In 1880 Captain Brewer was elected Sheriff of Chatham County, and re-elected four successive terms. He died 1 March, 1897.

Brave in battle, gentle in peace, charitable and honorable in all his dealings, beloved and respected by all who knew him, he was a model citizen, and has left a good name that his children can justly claim as their proudest heritage.

Captain Joseph R. Ballew, Company F, who became Captain of Company F on the promotion of Captain Rankin, as Major; was born 20 April, 1832, in Burke County. In 1852 he went to California via Charleston and Panama.

It required 130 days to make the trip. In 1859 he returned to North Carolina, making the return trip in 22 days. On the breaking out of the war, he was elected First Lieutenant of Company F, Twenty-sixth Regiment.

Captain Romulus Morrison Tuttle, Company F, famous as having commanded a company which at the battle of Gettysburg, out of 91 rank and file taken into action, had every man killed or wounded, himself among the number (wounded); was born in Lenoir, Caldwell County, N. C., 1 December, 1842, and left school in July, 1861, to join the army; was successively Orderly Sergeant, First Lieutenant, and Captain of Company F, Twenty-sixth North Carolina Regiment.

Was wounded four times in the four years service, viz: At Gettysburg, 1 July, 1863, right limb seriously fractured below the knee, which has never gotten entirely well; at the Wilderness, 5 May, 1864, centrally in the breast by minie ball, a flesh wound only—here his company lost 19 out of 26 men taken into action; west of Petersburg by a four-ounce canister ball in left breast, causing an ugly contusion and great suffering; and on 30 September, 1864, on the Squirrel Level road, south of Petersburg, in left forearm by minie ball, shattering the larger bone and necessitating a resection of three or four inches.

At the reorganization of the regiment for the war, April, 1862, Orderly Sergeant Tuttle was elected First Lieutenant, and on the resignation of his Captain, Jos. R. Ballew, in the Fall of the same year, he was promoted to the Captaincy.

After the war this battle scarred veteran, but mere youth in years, returned to college to complete his education, and in June, 1869, graduated at Davidson College, N. C.

He now (April, 1901) has charge of the Collierstown Presbyterian Church, near Lexington, Va.

Captain Henry Clay Albright, Company G. This gallant young officer, born 12 July, 1842, left college to enter the army as Second Lieutenant of Company G, Twenty-sixth North Carolina Regiment, and on Captain John R. Lane's promotion to be Lieutenant-Colonel of the regiment, Lieutenant Albright was made Captain of the company.

He was a "wonderfully good officer" is the testimony of his regimental commander. He passed unscathed through all engagements and battles, though present with his regiment all the time, until the spirited engagement of 29 September, 1864, on the Vaughan road, south of Petersburg, he was mortally wounded, and on 27 October he died in the Winder hospital.

Captain William Wilson, Company B, was killed at Gettysburg on the first day's fight gallantly leading his men up the hill and through McPherson's woods. Left school to join the army, and in June, 1861, was elected First Lieutenant of Company B, Twenty-sixth North Carolina Regiment. At the reorganization of the regiment in April, 1862, he was elected Captain. He would have achieved higher command had he survived the fateful battle of Gettysburg. He was buried on the field by the side of his Colonel. They were stricken about the same time and fell within a few feet of each other.

Captain William Pinckney Martin, Company H, was born 4 October, 1817. He was elected a delegate to the proposed Constitutional Convention 28 February, 1861; but as the calling of the Convention was defeated, he did not take his seat. His was the first company that volunteered from his county. It became Company H, Twenty-sixth Regiment. He was shot in the head at the battle of New Bern just before the regiment had orders to retreat, and was buried on the field.

Captain James D. McIver, Company H, was born in Moore County, N. C., 14 December, 1833; graduated from Davidson College in June, 1859; volunteered in the first company raised in his county, and was elected Second Lieutenant in July, 1861. This company became Company H, in the Twenty-sixth North Carolina Regiment.

On the resignation of Captain Clement Dowd in the Spring of 1862, Lieutenant McIver succeeded him as Captain of the company and remained in the regiment until the Fall of 1863; was in all the battles in which his regiment was engaged up to that time, except the battle of Gettysburg, at which time he was absent on furlough. Captain McIver was a most gallant and competent officer, and his leaving the regiment was much regretted. After the war he was County Solicitor, member of the Legislature in 1876, Solicitor of his District in 1878-1886, Judge Superior Court 1890-1898.

Captain James C. McLauchlin, Company K. This accomplished officer became Captain of his company in the reorganization for the war, April, 1862. He was wounded at Malvern Hill and again at Gettysburg, this last time so severely that it disabled him for service in the field, and he resigned from the regiment to accept lighter duty. Since the war for more than twenty years and at the present (April, 1901) Captain McLauchlin has been Clerk of the Superior Court for Anson, his native county.

Captain Thomas Lilly, Company K, who succeeded to the command of his Company, K, on the resignation of Captain McLauchlin, was also wounded at Gettysburg. He rose from Corporal and became recognized as one of the best officers in the brigade. He had command of the sharpshooters of the brigade, and fell mortally wounded 25 March, on the lines at Petersburg.

Lieutenant Orren Alston Hanner, Company E, enlisted 28 May, 1861, at the age of 18 as a private in Company E, Twenty-sixth North Carolina Regiment; was wounded at Malvern Hill 1 July, 1862; promoted to Second Lieutenant of the company in October, 1862; severely wounded at Gettysburg 1 July, 1863, and captured on the retreat of the Confederate army; carried first to hospital at David's Island, New York, then to prison at Johnson's Island, Ohio, where he remained until paroled in March, 1865. Has been a member of the General Assembly in 1872, 1874, and 1880.

Lieutenant Hanner was one of the bravest and best subaltern officers of the regiment. He and his Captain (S. W. Brewer) were both wounded and captured at Gettysburg, and the First Lieutenant, John B. Emerson, was mortally wounded at the same time. Captain Brewer's and Lieutenant Hanner's imprisonment prevented their being promoted to the positions of Major and First Lieutenant respectively.

First Lieutenant Gaston H. Broughton, Company D, was born in Wake County, 1838, enlisted in Company D, 1861, was promoted First Lieutenant 28 April, 1862, was wounded at the foot of the stone wall in the third day's charge at Gettysburg and remained a prisoner till the end of the war. He has been a farmer and a good citizen since the war and is now custodian of the Supreme Court building in Raleigh.

Lieutenant James G. M. Jones, Company D, was born near Holly Springs, Wake County, on 19 July, 1839. He enlisted in Company D, Twenty-sixth Regiment. At first a Sergeant, at the reorganization in April, 1862, he was elected Second Lieutenant of the company.

At Gettysburg, Lieutenant Jones was severely wounded in the hip. Through the kindness of Captain Young, Lieutenant Jones and his Captain (Adams) managed to get on a four-horse wagon loaded with wheat, and got safely to the Potomac river, and thence to the hospital at Richmond. He returned to duty 19 December, 1863, at Orange Court House, and took command of the company, his senior (Broughton) being prisoner of war, wounded and captured at Gettysburg. On 10 May, 1864, at Spottsylvania Court House, Lieutenant Jones was again wounded in the left breast, and would have been killed but for a daguerrotype of his sweetheart in his left breast pocket which deflected the ball. This lady he subsequently married. He returned to duty in September, 1864, and remained in command of his company until in the action at Burgess Mill, south of Petersburg, on 27 October, 1864, he was taken prisoner and confined at Fort Delaware until June, 1865, when he was liberated.

Lieutenant George Willcox, Company H, was born 17 June, 1835. He enlisted in Company H, Twenty-sixth North Carolina Regiment. At the reorganization of the regiment for the war in the Spring of 1862, he was elected Second Lieutenant of the company and remained as such until the Fall of 1864, when he was appointed Captain of Company H, in the Forty-sixth North Carolina Regiment, of Cooke's Brigade, in the same (Heth's) Division.

Captain Willcox was in all the battles and actions in which his command was engaged during the war, except at Malvern Hill, and when he was absent on wounded furlough. In the first day's fight at Gettysburg, he was badly wounded while carrying the flag of his regiment (see account of the battle in this sketch); was captured, but rescued on the retreat and returned to his command in time to take part in the battle of the Wilderness, in which battle he was again severely wounded, this time through the shoulder.

Returning to duty, he joined his regiment in the trenches around Petersburg, and was captured in the action at Burgess Mill 27 October, 1864, but escaped from the enemy during the night and rejoined his command. He represented Moore County in the Legislature of 1885-'86; also Moore and Randolph counties in the Senate in 1891-'92. Captain Willcox had three brothers in the war, he being the eldest. The next in age to him, W. M. Willcox, was a Lieutenant in Liddell's Brigade, Pat Cleburne's Division, General Bragg's army, and was killed at the battle of Chicamauga (September 19-20, 1863); Robert P. Willcox, another brother, was a member of Company H, Twenty-sixth North Carolina Regiment, and though several times wounded, survived the war several years. The youngest brother Herman Husband Willcox, as stated above, was killed at Gettysburg.

Lieutenant Wm. N. Snelling, Company D, enlisted on 10 June, 1861, in Company D, Twenty-sixth North Carolina Regiment. At the reorganization of the regiment, in the Spring of 1862, he was made Orderly Sergeant, and after Gettysburg, he was promoted to be Third Lieutenant. At this battle, every one of his company officers were killed or wounded, and Third Lieutenant Marion J. Woodall being killed, Sergeant Snelling was promoted Second Lieutenant, to date from 5 July, 1863, and placed in command of the company.

Lieutenant Snelling was twice wounded, once in the left breast and once in the leg. Except when recovering from these wounds, and once on a thirty days' furlough, Lieutenant Snelling was with his regiment, frequently detailed to act as Adjutant, and always ready for duty. He was with his regiment when it surrendered at Appomattox, and during the last few months of the war he was in command of Companies A. C and D, consolidated. Lieutenant Snelling made out the muster and pay rolls of his company from the beginning to the end, and would have received higher promotion, but from the fact that his Captain remained a prisoner of war after his capture at Gettysburg, and there was no vacancy.

Leonidas L. Polk, Sergeant-Major, was born in Anson County in 1837, and was of the same family as Colonel

Thomas Polk, President James K. Polk and Lieutenant-General (Bishop) Leonidas Polk. In 1860 he was a member of the Lower House of the General Assembly. In 1862 he enlisted in Company K, Twenty-sixth North Carolina Troops, and was soon appointed Sergeant-Major. In 1863 he was promoted to a Lieutenancy in the Forty-third North Carolina, and was severely wounded at Gettysburg. In 1864 he resigned upon being elected to the Legislature. In 1889-1892 he was president of *The National Farmers' Alliance* and died 11 June of the latter year and is buried in Oakwood Cemetery, Raleigh, N. C.

Private W. W. Edwards, Company E, was born 22 October, 1841, was in most of the battles in which the regiment was engaged; was wounded at Gettysburg, but returned to duty in time to take part in the battle of the Wilderness, May 1864, and the almost daily engagements with the enemy on the retreat to Richmond.

On one of these occasions, in front of the regiment was a school house occupied by the enemy's sharpshooters. It became necessary to drive them away and Colonel Lane called for volunteers for the dangerous work. Among those who responded was Private Edwards. Taking a few of his comrades with him, he crept up to the house and by a well directed fire, drove the enemy out of this house and the men were no more annoyed from that part of the line. After the war Mr. Edwards became associated in the publication of the *Messenger* at Siler City, and under the nom de plume of "Buck," became one of the most popular writers in the State.

THE END.

There is not a statement contained in this history that has not been obtained from official records, or from those who were actors in the events narrated. The mere recital of the story without embellishment is glory enough. Probably it will be vouchsafed to no soldiers in the future to suffer such a loss in open battle as the Twenty-sixth sustained at Gettysburg. There is no record in the past of such sustained heroism on a field of battle. Such being the case, it was meet and proper that the facts should be set out in detail; that honor should be given where honor was due. Such heroism as the Confederate soldier displayed cannot be in vain. Some good to the world must come from such sacrifice.

Nothing less than sublime confidence in the Justice of the Cause could inspire humanity to such deeds of glory, such endurance, such patriotism, and I close this history, paying this tribute to the private Confederate soldier, quoting the words of another:

"Let it be remembered there are other reasons than money or patriotism which induce men to risk life and limb in war. There is the love of glory and the expectation of honorable recognition; but the private in the ranks expects neither; his identity is merged in that of his regiment; to him, the regiment and its name is everything; he does not expect to see his own name appear upon the page of history, and is content with the proper recognition of the old command in which he fought. But he is jealous of the record of his regiment and demands credit for every shot it faced and every grave it filled.

"The bloody laurels for which a regiment contends will always be awarded to the one with the longest roll of honor. Scars are the true evidence of wounds, and regimental scars can be seen only in the record of the casualties."

"The men of the Twenty-sixth Regiment would dress on their colors in spite of the world."

In the preparation of this sketch, great assistance has been furnished by many of my surviving comrades and especially acknowledgment is due to Captain W. H. S. Burgwyn, Thirty-fifth North Carolina Troops, the brother of our lamented Colonel Harry Burgwyn. Captain Burgwyn is the historian of the Thirty-fifth Regiment, in which he served with great honor, and also of Clingman's Brigade, in which he later served with distinction as a staff officer. In the late Spanish War (1898) he showed he retained the military instincts of his family by again entering the service as Colonel of the Second North Carolina Regiment.

George C. Underwood.

Marley's Mills, N. C.,
9 April, 1901.

A

B

C

D

R

S

T

Y

Z

www.ingramcontent.com/pod-product-compliance
Lightning Source LLC
LaVergne TN
LVHW061245100826
845148LV00008B/1029
* 9 7 8 0 7 8 8 4 9 2 5 1 8 *